Superstrings

and the Search for The Theory of Everything

F. David Peat

CB
CONTEMPORARY
BOOKS
CHICAGO · NEW YORK

Library of Congress Cataloging-in-Publication Data

Peat, F. David, 1938–
Superstrings and the search for the theory of everything.

Bibliography: p.
Includes index.
1. Superstring theory. 2. Grand unification theories.
3. Twistor theory. I. Title.
QC794.6.S85P43 1989 539.7′2 89-15824
ISBN 0-8092-4257-5

Illustrations by Rosemary Morrisey-Herzberg

Figures 7-2 and 7-3 reproduced by permission of Roger Penrose

Published by Contemporary Books, Inc.
180 North Michigan Avenue, Chicago, Illinois 60601
Manufactured in the United States of America
Library of Congress Catalog Card Number: 88-23754
International Standard Book Number: 0-8092-4637-6

Published simultaneously in Canada by Beaverbooks, Ltd.
195 Allstate Parkway, Valleywood Business Park
Markham, Ontario L3R 4T8 Canada

This book is dedicated to the Resonances:

Eleanor

Matthew

Emma

Jason

Sarah

By the Same Author

A Question of Physics (with P. Buckley)
The Nuclear Book
The Looking-Glass Universe (with J. Briggs)
In Search of Nikola Tesla
The Armchair Guide to Murder and Detection
Artificial Intelligence: How Machines Think
Synchronicity: The Bridge Between Matter and Mind
Science, Order and Creativity (with D. Bohm)
Quantum Implications: Essays in Honour of David Bohm (co-editor with B. J. Hiley)

Acknowledgments

THE AUTHOR WOULD like to thank Professor Roger Penrose and his twistor group at Oxford University for their hospitality and for many interesting discussions. He would also like to thank Professor Chris Isham of the Imperial College of Science and Technology, London, and Professor Michael Green of Queen Mary College, London, for helpful discussions.

Thanks are also due to the researchers, including John Schwarz, David Gross, and Edward Witten, who were kind enough to send me copies of their preprints and answer telephone inquiries.

Roger Penrose and Michael Green were kind enough to read over parts of an earlier draft of this manuscript and offer a number of helpful comments and useful criticisms. Thanks are also due to Professor M. K. Sundaresan of Carleton University for helpful comments.

The author would also like to thank Harvey Plotnick and J. D. Fairbanks at Contemporary Books for their editorial insights and guidance.

Contents

Introduction 1
Chapter 1: A Crisis in Physics 9
Chapter 2: From Points to Strings 31
Chapter 3: Nambu's String Theory 57
Chapter 4: Grand Unification 71
Chapter 5: Superstrings 97
Chapter 6: Heterotic Strings: Two Dimensions in One! 133
Chapter 7: From Spinors to Twistors 163
Chapter 8: Twistor Space 201
Chapter 9: Twistor Gravity 237
Chapter 10: Into Deep Waters 275
Personal Postscript 339
Suggestions for Additional Reading ... 345
Glossary 349
Index 355

Introduction

IN THE SUMMER of 1984, two scientists sat down to work together at the Center for Physics in Aspen, Colorado. John Schwarz from the California Institute of Technology and Michael Green from Queen Mary College at the University of London were meeting yet again in an attempt to reach a breakthrough in a theory that had been plagued with paradoxes and anomalies. Over the last five years, the two men had been exchanging ideas and insights each summer and puzzling over a new approach to the quantum theory in which the elementary particles are treated as extended objects called strings. While during the first half of the 1970s John Schwarz had been working on an early version of string theory, he had since turned to other aspects of elementary particle physics. But Michael Green, a younger man, was still fascinated by the idea of strings.

Green and Schwarz had met by chance five years earlier when in 1979 they both happened to be visiting the European Center for Nuclear Research (CERN) in Switzerland. Over coffee one day the two physicists began to talk about their respective interests. Possibly that old theory of strings could be revitalized by an injection of some of the very latest ideas that were floating around theoretical physics at the time. Green and Schwarz therefore agreed to meet each summer to continue with their investigations.

The first string theory had its origins in a curious model of the heavier elementary particles that was created by Yoichiro Nambu in the late 1950s. (Nambu, a distinguished physicist in his own right, was teaching at the University of Chicago at the time.) The theory at first showed promise, but by the end of the 1970s it had become so riddled with contradictions that physicists were ignoring it in favor of alternative approaches like supersymmetry and the grand unified theory. Schwarz and Green were therefore free to explore this backwater of physics and claim it for their own. But progress was slow until, one day in August 1984, everything suddenly fell into place. A theoretical miracle had taken place, and *the theory of everything* was born.

Or at least that is how the story of superstrings is usually told, for by now the birth of this theory has reached mythic proportions within the scientific community. Of course, any scientific theory is built by the contribution of many hands. But in any event, a breakthrough certainly did happen, although considerable hard work was needed in addition to that liberal dose of intuition. Not only did this theory of Green and Schwarz work, but it did so with such an economy of means that it exhibited that measure of inevitability which is the hallmark of all great scientific ideas. Within days the two scientists had announced their results to their colleagues at Aspen, and only weeks later the international scientific world was claiming that superstrings are "the theory of everything." At last the problems that had plagued physics over the last decades were being swept away, and a revolution was truly taking place within the theory of subatomic matter. A single scheme, it appeared, was so powerful that it could explain not only the nature of the elementary particles but also the structure of space-time and even the very processes that occurred during the big bang origin of the universe itself.

The prestigious journal *Science* hailed the superstring revolution as "no less profound than the transition from real numbers to complex numbers in mathematics." In the pages of *Nature*, it became "a profound generalization" able to "unify gravity with the other fundamental forces in an almost unique manner." A number of popular articles followed. Michael Green himself wrote an account for *Scientific American*, while *Discover* ran the theory as its cover story. The theory was still making news in 1987 when the *New York Times Magazine* featured Edward Witten, one of the superstring superstars, in a cover profile.

Today the cheering has quieted down, in the scientific community at least, and superstrings have become a mainstream topic within elementary particle physics. Subatomic matter is pictured in terms of quantum strings that have an incredibly short length of 10^{-33} cm. Such a distance is very hard to imagine—even for a theoretical physicist! It means that if 1,000,000,000,000,000,000,000,000,000,000,000 of these superstrings were laid end to end, they would be just one centimeter long. It means that if we could imagine shrinking down from our own size to that of a single elementary particle, then we would then have to perform an equally powerful act of the imagination to shrink down to the size of a string. For everything that we have assumed to be elementary and ultimate about nature—the elementary particles—lie midway between the size of strings and the dimensions of our bodies. In other words, superstrings are incredibly short!

Not only do these strings account for the structure of the elementary particles, but they also provide a natural explanation for all the forces and interactions of nature—and possibly even for the underlying structure of spacetime as well. Each month dozens of new articles are published that explore the implications and ramifications of superstring theory, or suggest new variations and possible modifications. Superstrings (or at least some essential aspect of Schwarz and Green's initial insight) are here to stay, for the theory provides not only a new approach to the elementary particles but leads to something far deeper: a total revision of the mathematical language in which scientific reality is expressed.

While superstring theory was evolving from the early string model of the elementary particles, the English mathematician Roger Penrose was developing his own approach using new mathematical objects called twistors. We shall meet Roger Penrose in the later chapters of this book. Like strings, his twistors are extended objects that have the potential for describing both the internal structure of the elementary particles and the quantum nature of space-time. While Penrose's approach has quite different origins and is expressed in a different mathematical form, nevertheless there are arguments which suggest that a deeper theory may eventually combine elements from superstrings and twistors. Some conjectures on what this theory may look like are offered in the final chapter.

Until the advent of a theory of superstrings and of their cousins the twistors, quantum theory had been based upon the notion of continuous fields and a geometry which involves dimensionless points as its key feature. You may remember from school geometry that a point has location but not dimension. This means that points have no size at all and that an infinite number of these points can be placed on a line, or will fill a small region of space. Continuity means that space can be divided into smaller and smaller regions without limit, until these regions are infinitely small. These ideas of infinity, continuity, infinitely small regions, and dimensionless points are fundamental to the sorts of mathematics that twentieth-century physics has inherited. But this implies that these ideas of infinity and continuity are also a property of the quantum description of the world. As we shall see in the chapters that follow, there are deep reasons for wanting to get rid of these ideas, and it may well be that many of the difficulties that physics has been facing can be traced to these imported concepts, as it were.

Many physicists felt uncomfortable with the idea of a dimensionless point as the fundamental building block of space-time and quantum theory. There was already evidence that the elementary particles must have some form of internal structure and cannot be treated as points. But no successful theory had hitherto been able to treat them as extended. Now Schwarz and Green were showing that the most fundamental entities in the universe are not points but extended objects like strings that can spin, rotate, and vibrate. Similarly, with his twistor theory, Penrose was suggesting that space-time is built out of primitive, extended objects a little like light rays. In this latter approach, the elementary particles are formed out of conjunctions of these twistors, while geometric points are themselves secondary objects that arise out of the intersection of twistors. Clearly superstrings and twistors are pointing in the same direction.

Not only are the elementary particles a manifestation of superstrings, or twistors, but so also are the forces that act between them. Even space-time itself has its origins in the shimmering world of superstrings or the interweaving of twistors. These new approaches therefore hold out the possibility that relativity and quantum theory, the twin cornerstones of modern physics, will at last be unified and reconciled.

Superstrings offer a view of nature that is both elegant and economical. Whereas earlier approaches had been forced to make assumptions that involve grafting different features together in arbitrary ways, superstrings simply turn out to be the only possible theory of subatomic nature that can be constructed, in a consistent way, out of extended objects. It shows how nature can be described in terms of one of two possible symmetry patterns. (Scientists try to understand the elementary particles by grouping them together rather like animals in a zoo. Before superstrings came along, there were a host of alternative patterns, or symmetry groupings, with no way of choosing between them. One of the triumphs of superstrings is that only two contending symmetry patterns emerged uniquely out of the theory.) The theory also suggests that our universe may have evolved out of a higher dimensional space during the first instants of the big bang. It should also be able to explain the strengths of the four forces of nature, and it even indicates a universe of "shadow matter" that exists in parallel to our own.

Despite these achievements, it turns out that superstrings point even deeper, for they are forcing scientists and mathematicians to look into the heart of physics and change the mathematical language they have been using over the last 350 years. Its revolutionary implications may therefore extend beyond a theory of elementary particles and into new and even more subtle orders of description and fresh mathematical languages.

Twistors, for their part, were developed along a different route. Rather than requiring a higher dimensional space for their origins, they are defined in a three-dimensional space, but one whose dimensions must be described in terms of complex numbers. (We shall look at complex numbers, and the mathematics that follows from them, in Chapter 7. Essentially they are extensions of ordinary numbers—obtained by including $\sqrt{-1}$, also called *i*, the imaginary number. With the help of these complex numbers it is possible to extend mathematics in powerful ways.)

For Penrose this mathematical property of complexity is an essential feature of quantum theory and must be present in any description of space-time at the subatomic level. Working within this complex space, Penrose and his group have created some remarkable insights into the nature of the fields and forces of physics. Twistors also point to a physics that is already unified, one in which space-time and matter emerge naturally out of a single description.

These new scientific ideas are like those first buds that appear in spring; they must be given warmth and protection before they will unfold. A sudden critical frost can kill them off for good. It is in the unfolding of the theories of superstrings and twistors that new insights will occur. On the one hand promising new paths could turn out to be dead ends, while present difficulties may eventually lead to clues and further insights. It is difficult to predict how these new ideas will develop and eventually connect with each other. Who would have guessed that the abstract non-Euclidean geometries that were being investigated by Bernhard Riemann in the last century would turn out to be the exact mathematical backbone that Einstein required for his new theory of relativity? Or that the abstruse studies in functional analysis made by David Hilbert at the turn of the nineteenth century would yield the very Hilbert space needed for the new quantum theory?

Human creativity involves a free-flowing play of the mind in which new ideas constantly surface and interact with each other. Ideas are like patterns in a kaleidoscope which move and transform until some new pattern swings into perception. At the moment, physics is still reeling from the superstring revolution. Some physicists are already at work on extensions, moving in new directions, transforming, developing new mathematics. Others like Michael Green, however, caution that this vast mathematical formalism also calls for a much deeper foundation and for a profound new insight as to the meaning of the whole theory.

Here in 1988 it is as difficult to predict the physics of the twenty-first century as it would have been to foretell the vast superstructure of modern physics that was to emerge out of Ernest Rutherford's first atomic experiments at the end of the nineteenth century. Yet no matter how physics will develop, aspects of the essential insights of superstrings and twistors will probably be retained. Physics has taken a giant step forward.

1
A Crisis in Physics

PHYSICS TODAY IS faced with a series of questions that must be resolved before we can truly say that we have significantly advanced our understanding of the universe: How did the universe begin? What is the origin of time? What is the nature of matter? What is the ultimate meaning of physical laws?

But can such questions ever be answered in a final form? Some physicists believe that a complete description is indeed possible and that the fundamental questions will one day be answered. After that physics will become merely a matter of technological advances and fine tuning. Others hold that nature is inexhaustible and infinitely subtle, so that the dialogue between human consciousness and the universe will never end.

It could be said that the triumph of modern physics lies in its ability to pose such fundamental questions in rational, mathematical ways. That such questions have a formal meaning is a considerable achievement of the human mind. For while their origin can be traced to philosophical speculations formulated at the time of the Ancient Greeks, they have today become the burning issues of modern physics.

Over the last twenty years, fresh attempts were made to resolve some of these questions, and in so doing science was driven to construct theories like grand unification, supersymmetry, and supergravity. But, in the end, these approaches only served to highlight the magnitude of the difficulties faced by theoretical physics. While valuable insights were gained into, for example, the unification of electromagnetism and the weak nuclear force, a truly comprehensive theory of the elementary particles was still dominated by outstanding difficulties. In the end physicists in the late 1970s began to wonder if a radically new approach was needed, a totally fresh way of looking at the quantum nature of reality.

Then, in 1984, the theory of superstrings burst upon the world, and many physicists believed that this was the breakthrough they had been dreaming about. For, in addition to providing a new theory of the elementary particles, superstrings indicated that space-time and matter are not built on an underlying mathematical structure made out of dimensionless points but must be understood in terms of a new formal language. Extended objects called strings become the new foundation of geometry, force, and matter.

At the same time, the mathematician Roger Penrose was working, independently, on an alternative approach based on extended objects called twistors. We shall meet Roger Penrose and his work in Chapters 7, 8, and 9. While his twistor theory gives a different approach to understanding space-time and matter, it is possible that the ideas of twistors and superstrings may one day be united in a single great theory.

To understand the creative energy behind these rapid developments and why superstrings have generated so much excitement, it is a good idea to take a closer look at some fundamental questions concerning the elementary particles and their symmetries, the properties of space-time, and the failure of physics to reconcile relativity and quantum theory. Quantum theory is a theory about matter in the small scale and deals with elementary particles and atoms. By contrast, relativity is concerned with the structure of space-time in the large scale and with the way matter can curve this space-time to produce gravity. Although quantum theory and relativity deal with very different scales of things, there are deep reasons why these two theories should, at some level, be united. But, despite decades of effort, the two theories are still far apart. In addition to this difficulty, modern theoretical physics is faced with other puzzling questions. While superstrings or twistors in their present forms may not always be able to provide definitive answers, it was in grappling with some of these questions that these new theories were born.

What are the elementary particles?

In the first decades of this century, the elementary particles were thought of as the fundamental building blocks of nature. Molecules had been broken down into atoms, atoms into electrons and nuclei, and finally the nuclei into protons and neutrons. At this point, it was believed, the ultimate level of nature had been reached. Elementary particles are the primordial stuff of matter, and the entire universe is built out of vast numbers of these fundamental objects.

By the 1950s, however, the elementary particles had blossomed into a veritable zoo of some 200 different varieties. Faced with such diversity, the best that physicists could do was to group these particles into families according to a variety of patterns or symmetry groupings. But, in the end, trying to find an underlying reason for these patterns of the elementary particles only gave rise to a further series of questions.

There were also overwhelming problems involving quantum field theory, the mathematical framework in which the elementary particles are described. The approach was plagued with infinite numbers whenever important quantities were calculated. For example, the energy and the mass of an elementary particle should be very small, but explicit calculations gave infinite results. While it was possible to use a variety of mathematical tricks to get rid of these infinities—such as subtracting infinite terms from each other—the underlying problems of these infinities remained. The existence of these infinities caused some physicists to wonder if there was a basic flaw in the foundations of quantum field theory. Could these infinities somehow be related to the prevailing idea of infinite divisibility of space-time and the use of dimensionless points as the building blocks of geometry?

Before we leave the elementary particles, it is necessary to add a major qualification to the assumption that the analysis of matter always leads us to simpler and more fundamental levels, a belief that has been given free rein in twentieth-century physics. Complicated objects like typewriters, automobiles, grandfather clocks, and radios consist of simple elements, each doing its own job. When these cogs, levers, springs, resistors, and transistors are brought together in the proper way, cogs and springs can tell the time, while transistors, capacitors, and resistors can play music or calculate numbers.

Physics pursues this analogy and suggests that the universe itself can be broken down into simpler and smaller parts called elementary particles. Physics, it is hoped, will one day reach the ultimate level of nature in which everything can be described and from which the entire universe develops. This belief could be called *the quest for the ultimon.*

But this sort of analysis works best with mechanical objects, and there is every reason to believe that the universe is not mechanical. After all, such an approach is less successful when it comes to social and biological systems which have great internal complexity, include many feedback loops between their various levels, and have significant connections to the external environment.

In a similar way, it is possible that as we probe the elementary particles, superstrings, or twistors, they may not lead to a single most fundamental level, but rather they may open into a world of ever greater richness and subtlety. Instead of the end of physics being in sight, science may enter a new realm of complexity and new forms of order that contain, for example, feedback loops from much higher levels and even interconnections with the large-scale structure of the universe itself. Rather than being *the theory of everything*, superstrings may be the door to another universe.

But now back to our series of fundamental questions.

What is mass?

A fundamental theory of nature must give an account of the nature and origin of mass. It must explain why some elementary particles are heavier than others, and how these exact masses are related to the fundamental constants of the nature and to the origin and the structure of the universe itself.

In general relativity, mass and energy are responsible for the curvature of space-time. So a complete theory of mass must take account not only of quantum theory but also of general relativity and the geometry of space-time.

Is mass truly fundamental? For Penrose, in his early work on twistors, physics begins in a massless world in which everything moves at the speed of light. Mass enters as a derived notion, a by-product of the interaction of more fundamental, massless objects.

Why are there four forces in nature?

Nature employs four forces: gravity, electromagnetism, and the weak and strong nuclear forces. Gravity is the most familiar of these: it is the force that pulls us to the ground, holds the moon in its orbit, and creates the tides. Curiously enough, this ubiquitous force is the weakest of all. The reason why this weak gravity seems so important to us is that the body that attracts us—the earth—is so massive. But try to measure the gravitational attraction between two bricks, and you would need a very sophisticated laboratory to detect anything at all.* In fact, it is

---

*Another reason why gravity dominates the universe is that this force cannot be shielded. A positive electrical charge will attract negative charges around it, and from a distance, its effects will tend to average out to zero. But when a piece of matter gravitationally attracts other matter around it, the result is to increase this gravitational interaction.

very easy to overwhelm the force of gravity with the electromagnetic force. Draw a comb through your hair, and it will pick up tiny pieces of paper—the electrostatic charge on the comb is dominating the gravitational attraction of the whole earth!

Gravity and electromagnetism are forces of infinite range and dominate the large-scale world in which we live. But, at the scale of the atomic nucleus, two new very short-range forces make their appearance. One is called the strong nuclear force and operates between protons, neutrons, mesons, and the other inhabitants of the nucleus. The other force is called the weak nuclear force and is responsible for the way in which various radioactive nuclei break up and decay. Since the two nuclear forces are of such short range, we don't see them directly in our human-scaled world. However, they are vitally important, since they are responsible for such things as the ultimate stability of matter and the heat and light we get from the sun. The nuclear and the electromagnetic forces have been incorporated into the quantum theory. Gravity, however, still lies outside this scheme, for deep problems are connected with the quantization of gravity.

But why should there be four forces and not one, or ten, or fifty?* Why do these forces exhibit such a wide range of strengths, from the weakness of gravity to the strength of the strong nuclear force? The grand unified theory, as we shall see in Chapter 4, went some way toward answering these questions, for it suggested that the three strongest forces have a common origin and that, during the first instants of the creation of the universe, they were unified.

And how will the force of gravity enter into this unified scheme? Superstrings and twistors are each able to offer a new answer to this question.

---

*Some physicists have recently proposed the existence of a fifth force. This hypothetical new force is supposed to explain some anomalous experimental results on the gravity force, but at present the proposal is highly controversial.

What are the symmetries of the elementary particles?

In trying to make sense of the multitude of elementary particles, physicists grouped them into different families and showed how the members of each family are related by an underlying pattern or symmetry. While the word *symmetry* is used, it should not be confused with those more familiar symmetries of triangle, starfish, or flower. These latter symmetries are relationships within our ordinary space, but the symmetries of the elementary particles are more like mathematical symmetries, symmetries that must be written down in abstract spaces.

The point about these elementary particle "symmetries" is that mathematically they behave in a way similar to the spatial symmetries that are more familiar to us. But, of course, this raises important questions: What is the relationship of these abstract spaces to the space-time we live in? Are they simply mathematical devices, abstractions of the mind, or do they have some deeper connection to space-time? Could the spaces of the elementary particles have some essential relationship to our space-time? Superstrings and twistors both look as if they may be able to shed light on this mystery.

Determining the correct symmetry patterns has become one of the most active fields of particle physics. Yet, until superstrings came on the scene, there was a wide diversity of such schemes. Physicists would have preferred a single symmetry classification that emerges naturally and inevitably and that explains the relationship between abstract symmetry spaces and our own space-time. Superstrings have that inevitability.

Finally there is the issue of what physicists call "symmetry breaking." The grand unified schemes of particle physics are not actually observed in nature. According to the basic symmetry approach, particles that are grouped together should all have the same mass. In practice this does not happen, so that the basic pattern or symmetry is broken. But what sense does it have to group the particles together into symmetry patterns when, in actuality, they violate these patterns?

The contemporary approach is to switch attention from the particles as such to the underlying laws that give rise to them. Thus, while the *laws* of nature have a deep underlying symmetry, their actual solutions, the particles themselves, do not. The ideal symmetries of the laws are therefore broken or hidden by the real elementary particles. So while the elementary particles have their ultimate creation in a highly symmetric state, they must then pass through a series of symmetry breakings in which their masses are changed.

Roger Penrose, as we shall see, disagrees with this approach and argues that certain fundamental laws violate these symmetries at their origin. Rather than the laws being perfectly symmetric and their particular solutions, the elementary particles, breaking these symmetries, Penrose argues that even the basic laws of nature may be asymmetric. In Chapter 9 we shall see, for example, how the quantum of the gravitational field always appears in two forms, one that twists to the right and one that twists to the left. Penrose argues that at the very deepest level, these two forms are not equivalent, and nature is basically asymmetric. By contrast others rebut Penrose, saying that this asymmetry is not inherent in the deepest laws of nature but arises at some secondary level.

Why is our space three-dimensional?

With the addition of time, the space-time we live in is four-dimensional. But could a universe exist in two or five or even twenty dimensions? Is there a compelling reason why there are only three dimensions of space and one of time? This becomes a key issue in superstring theory, which is formulated in a higher-dimensional space. For his part, Roger Penrose suggests that, at the quantum level, nature begins in a space of three dimensions which must be described using complex numbers rather than our more usual real numbers. For Penrose, space at its most fundamental level requires the much richer complex mathematics that we shall first meet in

Chapter 7. Recently there have been some powerful mathematical discoveries which suggest that a four-dimensional space-time may be, in some ways, unique.

Why is the cosmological constant zero?

Einstein taught that space-time is curved and deformed by the presence of matter and energy. Whenever matter or energy is present, Einstein's equations allow physicists to calculate the new curved geometry of space-time. But the curious thing about these equations is that they permit for curvature even when no matter or energy is around. In other words, a curved space-time is a perfectly legitimate solution to Einstein's equations even in a totally empty universe. The amount of this zero mass curvature is given by what Einstein called the cosmological constant.

But accurate experiments indicate that, to an extremely high degree of accuracy, this cosmological constant must be zero. In other words, empty space does not curve. But why should this be so? What undiscovered principle of nature forbids any curvature in an empty universe? Is there some error in Einstein's basic equations? This remains a key question that superstrings may be able to answer.

What is the relation between relativity and quantum theory?

Modern physics is built on twin foundations—quantum theory and relativity. Yet despite half a century of hard work by some of the world's leading physicists, these two theories have stubbornly refused to be reconciled, but continue to coexist in paradoxical and incompatible ways.

On the one hand, relativity and quantum theory are irreconcilable; yet on the other, they are mutually dependent. General relativity is a theory about the structure of space-time, curved geometry being determined by the amount of energy and matter present. But matter and energy are quantum mechanical in nature, so a complete account of space-time geometry cannot ignore the quantum nature of the matter and energy which creates its very form.

The practical way of determining this geometry is to survey it using light rays and clocks. Even here on earth, a surveyor uses laser light to measure distances. The distances of planets and space probes are similarly calculated using radar or radio waves—employing a very accurate clock to determine the elapsed time of the signal. But light is quantum mechanical in nature, and the most accurate clocks are atomic clocks. To make precise measurements of space-time geometry, we must rely upon systems that are fundamentally quantum mechanical. Relativity again leans on quantum theory.

Likewise, the interpretation of quantum processes relies upon aspects of the relativistic world. To obtain a definite outcome to a quantum process, it is necessary to use experimental apparatus at the human scale. (Keeping everything at the subatomic level involves only probabilistic predictions with no definite outcome.) But since this human-scale apparatus comes within the domain of relativity, the experiments used to define quantum states must have a relativistic description.

Nuclear and electromagnetic forces are interpreted quantum mechanically as the exchange of a quantum force particle. The electromagnetic force between an electron and a proton, for example, is pictured as involving the interchange of photons—the whole thing is a little like two football players running down the field and tossing the ball, a photon, to each other.

But what about nature's fourth force, gravity, which holds the moon in its orbit through its attraction to the earth? There are strong arguments that it cannot be left out of the quantum theory. This means that gravity must also be pictured in terms of an exchange of the quantum particles called gravitons. But gravity is not just a force, it is also the curvature of space-time, for Einstein has taught us that what looks like the force of gravity is really the curvature of space-time. Now this is a very curious result, for it implies that, at the subatomic level, space-time curvature has to be explained in terms of quantum mechanical gravitons. Somehow the structure of space-time itself demands a quantum mechanical explanation.

Quantum processes are also related to space-time curvature in the immediate neighborhood of black holes. It was Stephen Hawking, the brilliant Cambridge physicist, who first argued that close to a black hole, the extreme degree of curvature of space-time actually creates elementary particles. Matter, according to Hawking, is created out of the fabric of space-time itself. But if geometry is the source of quantum matter, then clearly relativity and quantum theory have to be unified at some deeper level.

Toward a Unified Theory of Relativity and the Quantum World

The weight of the preceding arguments points to the need for a unified theory. Yet whenever scientists have attempted to bring relativity and quantum theory together they have failed. Some physicists have wondered if this basic incompatibility arises in the very definition of space-time itself. Indeed a closer analysis suggests that relativity incorporates a limited paradigm about space-time structure that stretches back for 300 years.

While Einstein revolutionized the Newtonian concept of space and time, he nevertheless continued to assume that space is continuous. That is, the properties of space continue unchanged to smaller and smaller scales: Space can be divided and subdivided right down to the dimensionless point, with all its properties changing smoothly from point to point. In mathematical terms, space-time and the theory of relativity use the language of calculus and differential equations, the basic grammar of science that has been employed since the eighteenth century. Could it be that Einstein's revolution did not go far enough?

Quantum theory shows that the infinite divisibility of space must be limited. The reason can be seen in Heisenberg's uncertainty principle, which states that the energy confined within smaller and smaller regions of space-time becomes increasingly uncertain. The reason is not too difficult to see. Suppose you try to measure the exact energy of a quantum system within a given short time interval. Heisenberg's principle dictates that the smaller this time interval, the more uncertain the energy. Put in another way, Heisenberg's principle allows for small regions of space-time to borrow quanta of energy—virtual particles and photons of energy if you like—provided that they are quickly paid back.

The process is a little like a credit card transaction: provided the money that you "borrow" is paid back fast enough, the interest never goes on your bill. Now suppose that at some time in the future you are allowed to borrow an unlimited amount of money provided that the more you borrow, the faster you pay it back. It would then be possible to become a millionaire—for a second at least. Likewise, in the case of space-time, the smaller the region, the larger the amount of energy that can be taken on loan and the faster it has to be paid back.

At the length scale of the elementary particles, these energy fluctuations do not pose a serious problem. But in much smaller regions, the borrowed energy can act to curve space-time. (In Einstein's theory, energy and matter both will distort the fabric of space-time.) Indeed, at around 10^{-33} cm, that is, 1/1,000,000,000,000,000,000,000,000,000,000,000 cm), this borrowed energy is large enough to curve space-time right around itself and, in effect, create mini black holes. Space-time loses its smooth, continuous nature and breaks apart into a violently fluctuating foam.

Heisenberg's uncertainty principle was essentially formulated to describe nature at the scale of atoms and elementary particles. To use it to talk about space-time foam requires an extrapolation that is as great as the change of scale between our own bodies and the elementary particles! Who knows that physics may not change radically within that region? However, most physicists seem perfectly happy to apply Heisenberg's principle right down to the domain of 10^{-33} cm, and at such small distances, it is clearly necessary to abandon a space-time that is built upon the notion of points and continuity. The theory of relativity at this scale of things demands a totally new—and yet to be discovered—mathematical formulation.

Yet despite this clear indication that space-time cannot be subdivided without limit, even quantum theory itself continues to make use of the mathematics of continuity. Erwin Schrödinger's wave equation, for example, is a differential equation, relating what is happening at one point to what happens at another point an infinitesimally short distance away. Likewise, its solution, the wave function, is defined at each infinitesimal point in space.

On the one hand, quantum theory denies continuity and the ultimate reality of the dimensionless point, yet on the other hand, quantum theory and relativity both continue to make use of such notions in their mathematical underpinnings. Clearly physicists are being forced toward a new intuition. What is called for is a totally new way of thinking. Many physicists believe that superstrings or twistors contain the seeds of this mathematical revolution.

A New Vision of Space

In creating his theory of relativity, Einstein was forced to return to Newton and to examine the Newtonian assumptions about the nature of space and time. But did Einstein go far enough? Are there still unexamined preconceptions that stand in the way of our answering the questions posed in this chapter?

While Newton's absolute space and time were abandoned, Einstein still retained the assumption of a continuous space based on dimensionless points. This required the use of the mathematical apparatus called the calculus, within the calculations of Einstein's theory. But the arguments of the previous section suggest that it is exactly at this point that the next revolution must begin, and the foundations of physics must be excavated even earlier than the time of Newton, to the mathematician and philosopher René Descartes.

When Newton wrote of seeing further by standing on the shoulders of giants, Descartes must have been in his mind. As a physicist, the Frenchman rejected the notions of space as a void and argued that all is matter, so that even the motion of the planets can be described by vortices in a background plenum,* a view strongly rejected by Newton. But since Descartes was also a mathematician, he was concerned with giving a measure and order to this plenum. By a stroke of genius, he hit upon a method to label or specify a point anywhere in space.** His invention was the coordinate grid and its system of coordinates.

With two axes we can draw a grid on this page and, in this way, specify any point using two coordinates.

Figure 1–1
The Cartesian coordinates of the point are (4,5).

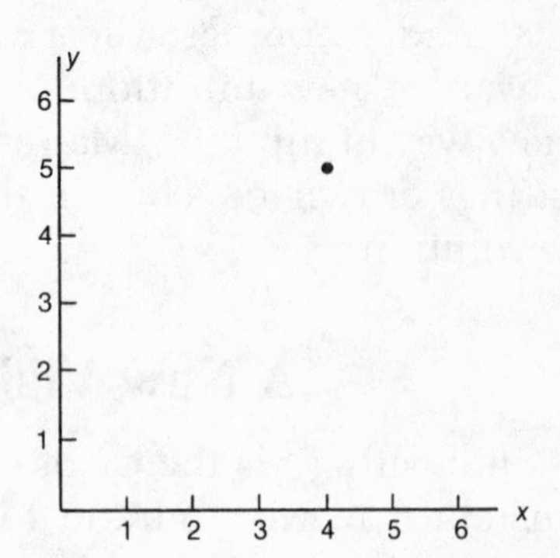

*A plenum is literally a full space, the opposite of a vacuum.

**In the third century B.C., Apollonius of Perga had tried to use a system of varying numbers in order to characterize the different shapes—circle, ellipse, parabola, and hyperbola—that are generated when a cone is cut into. But this system did not attract great attention and was never fully developed.

With three coordinates we can locate a point anywhere in three-dimensional space. Corresponding to each point, there is a trio of numbers; points and numbers become complementary ways of discussing space.

A triangle is defined by the positions of its three vertices, but these vertices now become three groups of coordinates. A line is a continuous set of points—a continuous set of coordinates.

Take the algebraic equation $y = 2x$, which allows y to be calculated from given values of x. If x is 1, then y is 2; if x is 2, then y is 4. Or, to put it another way, the pairs of numbers (1,2), (2,4), (3,6), (4,8), (5,10), . . . are all solutions to the equation $y = 2x$. But these pairs of numbers can also be thought of as coordinates in a plane. They define a set of points, and when these points are joined they produce a straight line. This line is therefore a representation of the algebraic equation $y = 2x$. Algebra is about numbers but it is also about points, curves, and shapes in space. *Algebra is about geometry, and geometry is about algebra.*

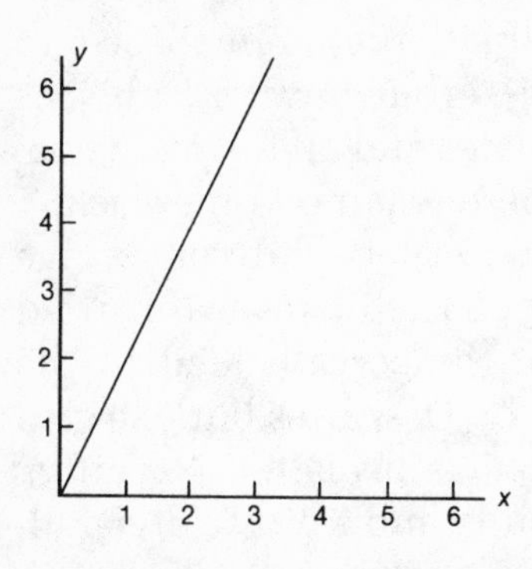

Figure 1–2
The equation $y = 2x$ has a geometrical representation as a straight line. Every point on this line has a pair of coordinates (x,y) which will satisfy the algebraic equation $y = 2x$.

At one stroke Descartes was able to link two great fields of mathematics. A line becomes an algebraic equation. The intersection of two lines becomes the solution of two simultaneous equations. Theorems in geometry and manipulations in space become transformations within algebra. The symmetries of the triangle, square, and pentagon are now properties of algebraic transformations.

Anything that can be done in geometry can be done in algebra. And why stop there—why stop at coordinates that contain just a trio of numbers? Why not have quartets, quintets, sextets, any number of numbers used as coordinates? In other words, why not have a multidimensional geometry?

Descartes had shown how to reduce geometry to algebra. And since much of physics is concerned with paths in space, spatial relationships, and symmetries, these also become problems in algebra. Suddenly much of physics is translated into algebra. Indeed one of Newton's great acts of genius was to extend this algebra by inventing the calculus. Calculus is a formal way of relating properties in space that are infinitesimally close to each other. By assuming that space is continuous, it is then possible to relate distant objects through an infinite but continuous series of these baby steps. Thanks to the calculus, Newton's laws could now be expressed in terms of algebra and what are known as differential equations. (Differential equations are ways of relating velocities, accelerations, and other rates of change that are determined at different [dimensionless] coordinate points in space and are assumed to be continuous.) From now on, physics was formally wedded to coordinates and all the mathematics that flowed from them. There was no going back.

While Newton's flat backdrop of space did not survive Einstein's revolution, the idea of coordinates was carried over. Admittedly they were no longer attached to the rigid rectangular grid proposed by Descartes but to a grid that could respond to the presence of matter. Nevertheless, the basic notions of continuity and infinite divisibility remained.

Even the quantum theory did not dispense with a space created out of dimensionless points. Schrödinger's wave function is a differential equation that uses essentially the same mathematics as that developed by Newton. The generalization of quantum theory, called quantum field theory, also relies on coordinates, for the quantum field is defined at each point in space and is a continuous function of Cartesian coordinates. The dimensionless point remains the basic paradigm of modern physics. Yet how can space-time be treated as infinitely divisible when Heisenberg's uncertainty principle sets a limit on such division?

Quantum field theory is plagued with such problems as the infinite results that are found when certain properties are calculated. Some physicists feel that the origin of these problems lies in the assumption that the quantum field is defined right down to infinitesimally short distances. Quantum theory seems to be demanding a new mathematical language for space-time.

Beyond the Space-Time Point

Coordinates are used as map references, to program the movements of industrial robots, to play video games, and to direct a space flight. We learn about them in high school, they are the basic grammar of the physicist, they pervade so much of our formal thinking. How will it be possible to do without them?

In the 1950s David Bohm argued that quantum theory is leading us toward descriptions of space that are topological rather than geometrical. Topology is concerned with boundaries, intersections, and containment. It is more general than geometry and could be thought of as those relationships which survive being stretched, bent, and twisted on a rubber sheet. Once lengths have been expanded or contracted, straight lines twisted into curves, and triangles transformed into circles or squares, all that remains are the more primitive, yet powerful relationships of topology. These are concerned with the way figures intersect and enclose each other and with how many holes pass through a figure.

Bohm felt that topological relationships, rather than coordinate grids, should form the basis of a new order that would be in harmony with the quantum theory. In-

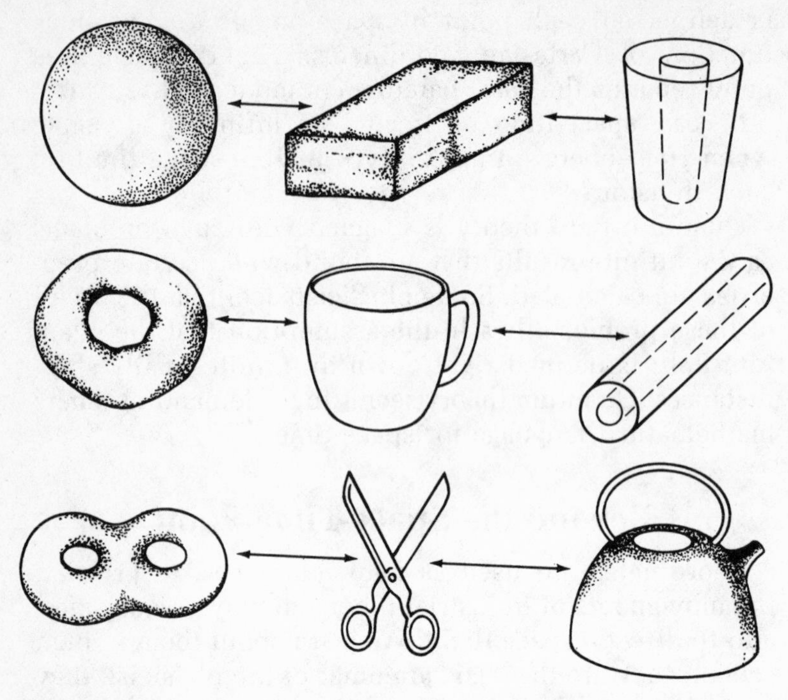

Figure 1–3
By stretching and deforming, it is possible to turn a beach ball into a cube or a beaker, but not into a doughnut. Sphere, ball, cube, and beaker are all topologically equivalent. Figures having a single hole are all topologically equivalent to the doughnut-shaped torus. Another class contains two holes, and so on.

deed, with new forms of topology it should be possible to abandon the ideas of space-time continuity altogether. During the following decade, Bohm pushed ahead with these ideas and developed an even more general kind of order, called the *implicate order.* The implicate order could be thought of as a fundamental description that is much closer to the spirit of the quantum theory. It suggests that the explicit patterns and relationships we see around us may not be a manifestation of nature at its deepest level. Rather they are the explicate forms of an underlying implicate or enfolded order. This implicate order has been compared to a holographic image in which information from many different parts of a scene is enfolded in a complex way onto the photographic plate. Within the implicate order, for example, it is possible for distant events to be directly connected.

A discussion of this implicate (or enfolded) order would take us beyond the topic of this present book,* but one idea of this approach is that this implicate order could underlie space-time and matter so that objects which are normally thought of as being separate and distant become interlocking and mutually contained within each other, and that movement of quantum objects would be seen in an entirely new light.

Roger Penrose was also puzzling about the continuity of space and asking what it means for there to be as many points between these two dots · · as within the whole universe. The idea seemed nonsensical to Penrose, who began to work on new ideas of space in which points are not the primary objects. At first Penrose investigated the implications of the *spinor*, which describes quantum mechanical spin. Spinors are curious objects, for they express the essential two-valuedness of the spin of elementary particles like the electron and proton. They are therefore essentially binary objects that combine according to simple arithmetic rules. Nevertheless Penrose was able to use them to create certain aspects of space. In other words, space, at the quantum level, appears to emerge out of very simple rules for combining quantum objects. But the spinor was not sufficiently general, and Penrose then went on to develop the twistor—the fundamental object of a new geometry of extended objects. We shall learn about this new twistor space in Chapters 7, 8, and 9.

Other researchers worked on alternative approaches.

---

*Bohm's ideas on the implicate order can be found in David Bohm, *Wholeness and the Implicate Order*, London: Routledge and Kegan Paul, 1980, and in David Bohm and F. David Peat, *Science, Order and Creativity*, New York: Bantam New Age Books, 1987.

Some suggested that there must be a minimum possible distance in space, or a basic uncertainty in the distance between objects, or that space-time has an underlying lattice structure. David Finkelstein attempted to derive the structure of space-time using the operations of symbolic logic. He had earlier been struck by the thought that the human brain is able to generate the concepts of space and time out of the firings of its neurons. Could nature itself have created space out of a similar network of logical relations?* The German physicist and philosopher Karl von Weisacker argued that quantum theory places time in a special position and suggested that space can be generated from what he called a "temporal logic."

Many such ideas were floating around in the late 1960s, suggestive accounts for a new physics, which would require considerable energy for their development. But by the mid-1970s physics was making great strides with supersymmetries and unified theories, and these other attempts to reformulate the structure of space-time never entered the mainstream of physics.

Two new solutions to this central problem of modern physics—what is the ultimate nature of space-time and matter—form the topic of this book. One of these solutions, superstrings, proposes that the elementary particles cannot be reduced to featureless points but are essentially one-dimensional objects, strings that rotate and vibrate. Not only does matter find its origins in these one-dimensional objects of incredibly minute size, but the strings themselves are irreducibly linked to an underlying space. Indeed space-time itself must at some level be created out of superstrings. The theory of superstrings therefore

*Very recently David Finkelstein made what he considers to be a breakthrough in his understanding of the origin of space-time and the quantum theory. His approach is analogous to the neural nets that are currently being investigated by Artificial Intelligence researchers, and in some ways recall the "spin networks" of Roger Penrose that are discussed in Chapter 7.

appears to describe all that is essential in physics—space-time, force, and matter. For this reason, it has been called the theory of everything.

The other approach discussed in this book is based on another kind of one-dimensional object: twistors. Twistors have their origin in a space of complex dimensions whose rich properties combine essential features of relativity and quantum theory. Twistors have also led to profound new intuitions in physics, and some theoreticians now believe that these twistors may be the proper starting point for the development of superstrings.

2
From Points to Strings

DURING THE FIRST years of this century, everything in the garden of physics was coming up roses. At the Cavendish Laboratories in Cambridge, the great Ernest Rutherford was shooting alpha particles (the charged nuclei of helium atoms) at a gold foil in an attempt to learn about the atoms out of which the foil was built. Theory at that time taught that atoms are like Christmas puddings, with the rich fruit of electrons and protons dotted inside at

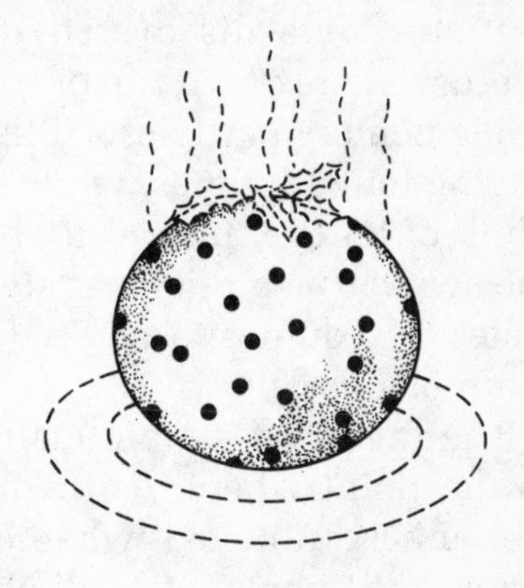

Figure 2–1
Rutherford's experiments demolished the early atomic theory in which protons and electrons are dotted about the atom like fruit in a Christmas pudding.

random. But when, to his surprise, some of these alpha particles were shot directly back toward him, Rutherford was forced to conclude that the atom must have a hard center. The atom was more like a miniature solar system than a Christmas pudding, with a tiny central nucleus for a sun, surrounded by a much less dense cloud of electrons. Within that nucleus, protons and neutrons were later to be found.

Physics seemed to have solved the great question about the ultimate nature of matter. All matter was composed of atoms and, in turn, these atoms were built out of three elementary particles—electron, proton, and neutron. Or there were only two if the neutron turned out to be a composite particle made from an electron and proton.

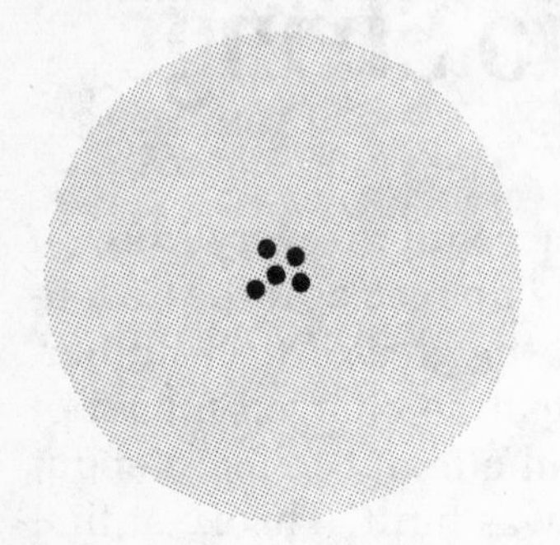

Figure 2–2
Rutherford's scattering experiments suggest that a tiny, dense nucleus is surrounded by a cloud of electrons.

However, as time went by, this simple picture became increasingly complicated. More and more elementary particles were discovered until, by the middle of this century, there was a zoo of some 200 different inhabitants or elementary particles. Of these some were well defined, having long lifetimes, while others existed for the merest blink of an eye. Indeed it was not always clear whether they were true particles at all or simply temporary excited states or composite states (also called resonances, but more of this later).

The theory of atomic matter had become complicated to the point where distinguished scientists like Werner Heisenberg doubted whether the elementary particles were fundamental after all and were not simply the manifestation of a deeper principle, related perhaps to fundamental symmetries.

The ordered garden of physics had been turned into an unruly zoo. But as in any zoo, it was still possible to classify the inhabitants. Just as the denizens of a zoo can be grouped together as mammals or reptiles, aquatic or land animals, those found in temperate and those in tropical climates, so too the elementary particles were gathered into families. It was by studying the relationships within and between these families that physicists hoped to discover clues as to the deeper, underlying order of subatomic matter.

One of the simplest ways of classifying the elementary particles is according to their spin. To give a particularly crude picture, the elementary particles spin like tiny planets. But being quantum mechanical in nature, these spins must have discrete, quantized values. It turns out that the different elementary particles can have a spin taken from one of the following values: 0, $\frac{1}{2}$, 1, $\frac{3}{2}$, 2, . . . ; that is, spin increases in units of $\frac{1}{2}$, which means that while some particles have fractional spins like $\frac{1}{2}$ and $\frac{3}{2}$, others have whole spins like 0, 1, and 2. In fact, it is useful to divide the particles into those with a fractional spin (like the $\frac{1}{2}$ spin of the electron, proton, and neutron), called *fermions,* and those with whole spins (like the photon with spin 1), called *bosons.*

This division into fermions and bosons is important for several reasons. To begin with, bosons and fermions obey different rules of encounter, or what physicists call different statistics. In concrete terms, fermions tend to keep away from each other and have different states or wave functions, while bosons are quite happy to congregate in the same state. This fermionic antisocial behavior is what prevents normal matter from condensing down into a featureless blob. Bosons, however, can gather together in a cooperative fashion and create the intense light of a laser or the superfluid behavior of liquid helium.

The classification into fermions and bosons is relevant in another way. Fermions make up the material side of the universe, while the bosons are responsible for its forces. Physicists call bosons the "carriers" of the weak, strong, and electromagnetic forces and, it is hypothesized, of the gravitational force. Therefore, at the quantum level, the forces of nature take the form of boson particles being exchanged between the fermions. (You will recall the two football players who toss a ball to each other as they run up the field.)

The division of particles into bosons and fermions seemed perfectly natural, but its deeper meaning was not fully understood at the time. It was only with the advent of supersymmetry in the 1970s that physicists learned, theoretically at least, how to turn bosons into fermions and vice versa. As a result, new insight was gained into the meaning of this particular zoological order. Supersymmetry was also to be important in formulating a correct superstring theory.

Another important classification of the elementary particles is according to the strength with which they interact. The four forces of nature are, as was explained in the previous chapter, gravitational, electromagnetic, and the weak and strong nuclear forces. For the moment we will forget about the first two forces, gravity and electromagnetism, and concentrate on the two nuclear forces. In fact, the strength of these forces provides a useful way of classifying particles, since not all of them experience the strong nuclear force. One group—containing the electron, mu mesons or muons, taus, neutrinos, and other fermions—is called the *leptons*, and interacts using only the weak nuclear force and the electromagnetic force. Leptons never feel the strong nuclear force. By contrast the *hadrons*—which include the proton, neutron, hyperons, and mesons—interact via the strong nuclear force as well as by the other forces of nature. It turns out that the hadrons are all much heavier than the leptons, a fact which suggests that their mass must somehow be related to the strength by which they interact. Since the leptons never experience the strong nuclear force, they are much lighter.

Within the division of leptons and hadrons, there are, as we shall see, yet further classification schemes and family groupings. The development of various family groupings, or particle classification schemes, convinced physicists that a series of deeper relationships must exist between the different particles in the family group. One explanation for these relationships is that the elementary particles are not elementary at all, but are composed of more fundamental entities called *quarks*.

In fact, this quark theory turned out to be an excellent way of understanding the elementary particles. If the elementary particles are created out of a few quarks, then we should indeed expect to see distinct family resemblance between the various particles. The effect would be similar to the relationships between the various colors on a painter's palette, which can be broken down into just a few primary colors.

At first the quark picture was received with great enthusiasm, and physics made some striking advances. But, in the long run, it turned out that none of these attempts at family grouping was entirely successful. There was, at times, something arbitrary about the particular choice for a family grouping or the way in which physicists decided which cousin should be placed within which family and how large to make the family group. Indeed it was out of a general dissatisfaction with this whole attempt to unify particle physics that superstrings were born and hailed as the approach that finally promised to bring order into the staggering complexity of the elementary particles.

Rutherford's Black Box

But we have moved ahead too rapidly. Let us return for a moment to Ernest Rutherford and his early scattering experiments. To Rutherford, the atom was a mystery. Far too small to be seen, it could not be taken apart or touched in any direct way. Its secrets were sealed inside its tiny space. How were they to be unlocked?

The problem faced by Rutherford can be illustrated by a science fiction story. Suppose that space travelers from earth discover a mysterious black box on Mars. They cannot open it, and x-rays fail to penetrate its interior. What, they wonder, does it contain? What is its function?

There are, however, a number of terminals on this box, and scientists guess that they are used to make electrical connections. In their first tentative experiments, the scientists send electrical signals into the box, and with meters and oscilloscopes, they measure what comes out. They soon discover that their input signals generate a systematic response within the box. By relating input to output, they begin to puzzle out the electrical components that are inside. Finally, without ever having been able to force the box open, they are able to draw a circuit diagram and deduce the box's function and purpose.

In a similar way, the atom was a black box for Rutherford. By firing alpha particles at the atom and observing how they were scattered by its interior, he was able to deduce that it had an outer cloud of electrons and a very minute but very dense inner core. The army of elementary particle physicists who followed Rutherford with their cyclotrons, synchrotrons, linear accelerators, colliding-beam storage rings, and the whole gamut of modern elementary particle accelerators, wholeheartedly adopted the great man's approach. First it was the atom that was viewed as a black box, then the nucleus, and finally the elementary particles themselves.

Modern particle accelerators take electrons, protons, or alpha particles and accelerate them to extremely high speeds—close to that of light—and then direct them at the modern black box—a target elementary particle. In a typical scattering experiment, physicists measure what is shot into the region of the elementary particle and what emerges. By relating what goes in to what comes out, they hope, like Rutherford before them, to discover the secrets of the elementary particles.

In a typical experiment, particle A is speeded up until it hits particle B. A and B could be thought of as the inputs into this scattering experiment. If elementary particles held no secrets, then this collision would be similar to what happens when two billiard balls collide. After their collision, *A* and *B* would fly off at different angles and different speeds. By measuring the speeds and angles, scientists can work out the strength of the interactions that occurred during the collision. Something analogous occurs with elementary particles during low energy collisions.

But when considerably more energy is involved in a collision, when the colliding particles are moving faster, the picture is radically different. Einstein had shown that energy can be converted into mass:

$$E = mc^2$$

It is therefore possible that the energy *E* released in a collision will be sufficient to create a new particle with a mass *m*. This particle, call it *X*, may be stable, or it may be short-lived so that it rapidly decays back into *A* and *B*, or it may decay into new particles *Y* and *Z*.

Already things are looking more complex, for there can be a variety of outcomes, or "outputs," to each collision. Quantum theory allows physicists to calculate the probabilities corresponding to each of these particular outcomes, which can then be checked against the results of particle collision experiments.

The Scattering Matrix

As physicists pushed their experiments to ever higher energies, the results became ever more complex, for new particles were constantly being detected. That black box called an elementary particle seemed to hold a variety of mysteries. To keep track of the experimental results in a formal way, John Wheeler in 1937 introduced what he called a scattering matrix, or S-matrix. This S-matrix could be thought of as containing the essential relationships of the black box for elementary particle scattering experiments. Since it is sometimes too difficult to calculate exactly what happens when two elementary particles interact, it is better to relate the free particles as they speed in toward each other to the resulting particles that speed out again.

To learn about superstrings, we don't really have to know much about the S-matrix beyond realizing that it is a mathematical way of relating all the incoming elementary particles to the outgoing ones, along with their excitations and resonances; in addition, it also gives information on the relative probabilities of these various processes. The S-matrix could therefore be thought of as a set of relationships between various quantum states. In turn, these relationships are the manifestations of the interactions and quantum processes that take place during the collision of elementary particles. In short, the S-matrix contains all the significant information that physicists discover during their experiments with the black box elementary particles. All that remains for them to do is to interpret this information correctly.

Physicists of that period were now faced with the difficult task of calculating all the possible results of a scattering experiment, along with the relevant probabilities, for the S-matrix and then comparing these with the actual results of the experiment. A significant step in speeding up this whole process was taken by John Wheeler's student Richard Feynman. Feynman discovered that the extremely complicated expressions involved in S-matrix calculations could be broken down into a series (an infinite series as it turned out) of much simpler terms. Each of these terms could, in turn, be represented by a diagram. These Feynman diagrams, as they are called, are pictorial representations of each term in the interaction series and correspond to rigorous mathematical expression.

Feynman diagrams can be classified, for example, according to how many loops they contain. Feynman showed that it was possible in a single step to add an infinite series of diagrams of a particular class. In other words, complicated quantum calculations could now be broken down into an infinite number of simpler terms,

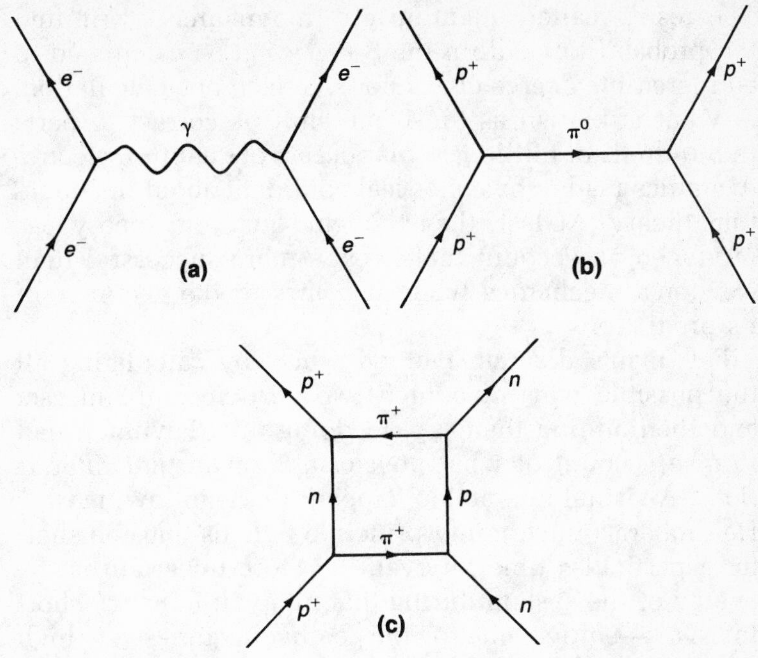

Figure 2–3
Feynman diagrams showing (a) two electrons interacting electromagnetically by exchanging a photon, (b) two protons interacting via the strong nuclear force and interchanging a neutral pi meson, and (c) a neutron and proton interacting with the exchange of two charged mesons.

which could then be added together.* We shall also see related diagrams popping up again in superstring theory, but in a new and curiously transformed way.

In the case of electromagnetic interactions, physicists were able to use Feynman diagrams to calculate exactly what happens when two particles collide and interact.

*In fact, these Feynman diagrams could be thought of as a series of approximations that, taken together, add up to the total interaction. Such an approach, in which a complicated quantity is calculated by adding up a series of approximations, is called perturbation theory and is often used in physics.

The result, called quantum electrodynamics, permitted the probabilities within the S-matrix to be calculated to an incredible degree of accuracy. In fact, one calculation, of what is known as the Lamb shift, is correct to parts in hundreds of millions. This success of quantum electrodynamics made physicists feel confident about the quantum theory. At last, they thought, quantum theory has produced an account that is even more successful than Newton's mechanics when it comes to the accuracy of its predictions.

Feynman's diagram method works by calculating all the possible ways in which two particles can interact and then adding them up. In doing this, Feynman had to take account of what physicists term *virtual interactions*. A virtual interaction happens in the following way: Heisenberg's uncertainty principle tells us that the shorter a particular time interval, the more uncertain is the energy of the system during this interval. For very short intervals—equivalent to the collision times of high energy particles—this energy uncertainty may be very large, large enough to create a new particle. In other words, new particles are brought into existence for short time periods, only to die back again and release their energy. It is as if quantum theory allows nature to borrow energy and then pay it back, to create and then destroy particles, provided that the sum of energy balances at the end of the collision. It is these so-called virtual particles that play the key role in interactions.

Feynman had to take all possible interactions and then add them together. In fact, there turn out to be an infinite number of such interactions, an infinite number of ways in which virtual particles can be borrowed from the quantum background and then paid back. But, since each term was smaller than the one that came before, it was still possible to add the series and produce a finite result. This is a little like summing up the series

$$1 + \frac{1}{2^2} + \frac{1}{3^2} + \frac{1}{4^2} + \frac{1}{5^2} + \ldots$$

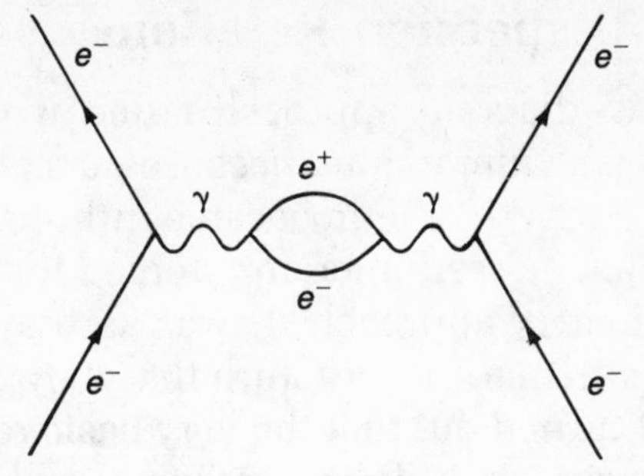

Figure 2–4
Two electrons interact via the exchange of a photon, but this time the photon itself creates an electron/positron virtual pair.

and getting a finite answer:

$$\frac{\pi^2}{6}$$

As far as the electromagnetic interactions were concerned, those involving collisions between leptons and leptons (e.g., electrons and positrons) or leptons and hadrons (e.g., electrons and protons), the method worked perfectly. Calculations of the S-matrix using Feynman diagrams agreed perfectly with experimental results. The real difficulties were encountered when scientists went on to calculate hadron-hadron collisions in which the strong interaction comes into play.

The problem with hadron-hadron collisions is that the strong interaction is simply *too* strong. When the first interaction term is calculated, using a Feynman diagram, everything seems reasonable, but successive terms do not turn out to be smaller; they are all more or less of the same size. Since the final result involves adding up an infinite number of finite terms, the sum blows up to infinity. The S-matrix cannot be calculated. But if the S-matrix blows up, how are physicists to make progress in understanding the inner nature of the strongly interacting hadrons? Physics had been blocked at the very next step.

Dispersion Relations

The Feynman diagram approach failed when it came to hadron-hadron interactions because successive terms were so large so that their sum became infinite. Physicists and mathematicians were therefore forced to take a fresh look at the S-matrix approach. It was as they started to examine its mathematical structure that they realized all was not lost. It turned out that the very basic requirement that the S-matrix should be mathematically well behaved—or, in mathematician's terms, that it should be analytic—imposed a powerful restriction on just how the matrix can be written down. There were also other requirements to do with the basic laws of causality and relativity. Taken together, these restrictions have the effect of giving structure to the S-matrix and producing internal relationships among its various parts. In other words, if one part of the matrix is known, then mathematical good behavior makes it possible to know about another part. Even if the matrix could not be calculated from first principles, using Feynman diagrams, it was still possible to make use of very powerful internal relationships.

Scientists call these relationships dispersion relations. In fact the term *dispersion* refers to the way in which a prism breaks up a ray of light into its various colors. Scientists had earlier discovered that the equations that describe this breaking up, or dispersion, of light are mathematically related to those which govern the way light is absorbed within the prism. On the grounds of good mathematical behavior and without having any theory about the prism itself, it is possible to relate the process of dispersion to that of absorption. In a similar way, physicists used the dispersion relations of the S-matrix to equate experimental results on scattering to those concerned with the appearance of short lived particles and resonances.

Dispersion relations deepen the mystery of subatomic matter, for they appear to be telling us that if we understand the details of how certain particles scatter off each other, then we also know something about the structure of other particles. In other words, taken as a whole, the elementary particles are self-referential; they pull each other up by their own bootstraps, as it were. At least this was the extreme position suggested by Geoffrey Chew. The whole elementary particle zoo, Chew argued, comes into existence by pulling itself up by its own bootstraps. No particle is more fundamental than any other particle; each is the product of all the others.

This was certainly an intriguing and self-consistent suggestion, and for a time, bootstrap theory became fashionable. But physicists still felt that something deep was hidden within the S-matrix. The various scatterings, resonances, and excited states that were found in elementary particle experiments must in turn be the result of an underlying theory of some more fundamental entities.

Regge Trajectories

At the cusp of the 1950s, an Italian physicist named Tulio Regge was working on certain mathematical properties of the S-matrix. In particular, he was interested in what are known as resonances, very short-lived phenomena that occur during a scattering experiment. In some cases the collision of two high-energy particles produces long-lived particles, but in others the outcome is less clear. The whole system seems to "ring" with energy like a bell, or, to put it another way, for a short time energy is localized in space, bound together within the inner region of the collision until it eventually decays away. These resonances behave like temporary localizations of matter and energy; they are like new particles with very short lives.

There was some debate as to whether the resonances that Regge was studying were in fact new particles or simply a temporary "ringing" of energy. What was particularly significant was that these resonances were extended in space. They also possessed a very large value for their spin. While the hadron particles normally have small spin quantum numbers, like $\frac{1}{2}$, some of these resonances had very large spins indeed. And particles with high spin could be said to have a sort of angular momentum, a property that is normally associated with a spinning ball of finite size rather than with a point particle.

In fact, Regge himself was concerned purely with the mathematical description of scattering processes and not so much with experimental details. But other physicists, inspired by Regge's work, began to study these resonances, trying to discover rules that would distinguish them from very short-lived particles and working out ways of relating them together. The problem was that there were so many of these hadron resonances that they couldn't all be fundamental. Regge himself felt that it should be possible to order these various "ringings" of the S-matrix. He found, for example, that he could plot a simple line on a graph that would link a family of resonances together. This plot is called a *Regge trajectory*.

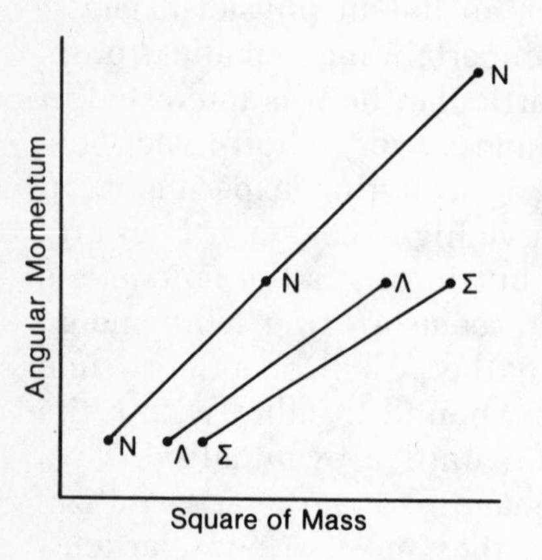

Figure 2–5
A series of Regge trajectories which show how some of the elementary particles, points on these trajectories, are related. *N* stands for either a proton or neutron.

The Regge trajectories also confirmed that resonances have angular momentum. In other words, they referred to spinning blobs of matter—extended objects or temporary particles that are smeared out over a finite region of space.

There was much debate about the meaning of these Regge resonances. Could all elementary particles be treated in this way? Are the elementary particles really extended objects with definite sizes? Louis de Broglie, one of the founding fathers of quantum theory, along with David Bohm and Jean-Pierre Vigier, had even attempted to work out a theory of extended quantum particles. But this approach, and others like it, ran into serious difficulties. There did not seem to be any consistent way of treating quantum objects as tiny balls or diffuse blobs while at the same time conforming to the dictates of relativity.

Dual Resonance

The next significant advance came from Gabriele Veneziano, who in 1968 showed how to relate the various resonances, collisions and scattering processes, and Regge trajectories within a single model. Veneziano wrote down a relatively simple formula which, he claimed, would explain all the essential features of the S-matrix, including the multiplicity of resonances, along with their different spin quantum numbers. He called his approach the dual resonance model. While it could not give a detailed account of the internal structure of the hadrons—the strongly interacting elementary particles—it did relate Regge's resonances to the scattering and exchange of particles in other experiments.

In essence, Veneziano's formula was making an assumption about the S-matrix as a black box. He had noticed that there were two sets of inputs into this black box—two scattering channels, *s* and *t* as they are called—which give "dual" descriptions of the same physics. The model seemed to be saying that what looked like a resonance, a ringing of energy, in one channel would from the other, dual channel look like an interaction. Veneziano had shown how to relate interactions—that is, the exchanges that occur between two points in space-time—to resonances.

Now let us look a little closer at what happens in the s-channel. In a typical elementary particle experiment, two particles, 1 and 2, come rushing toward each other. As they collide, they merge for a time to form a new collective mode called a resonance, *R*. This resonance persists for a time but finally breaks apart into new particles 3 and 4.

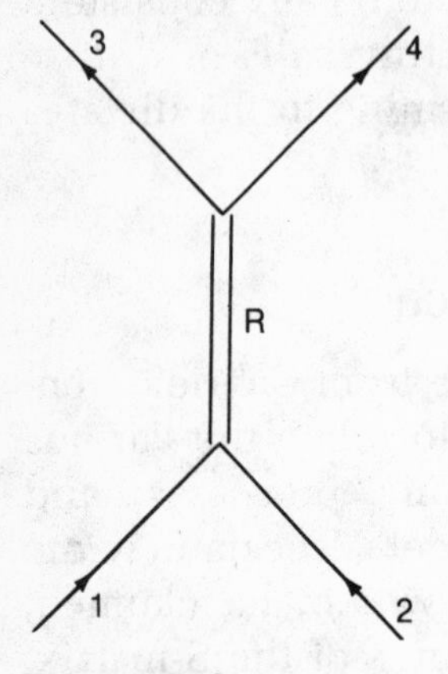

Figure 2–6
Particles 1 and 2 combine to form a temporary resonance *R*, which then disintegrates into particles 3 and 4.

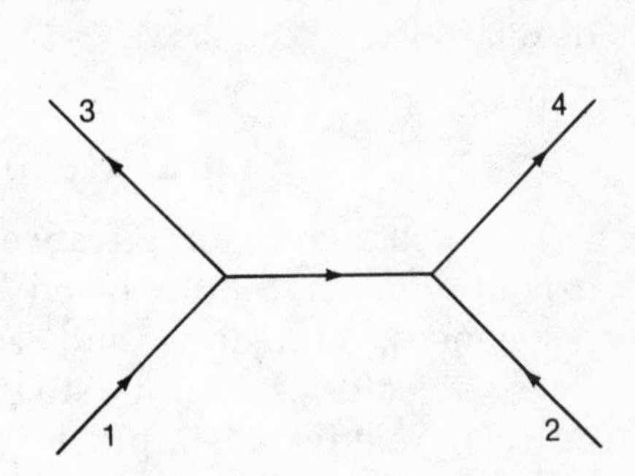

Figure 2–7
In this case, 1 and 2 interact via the exchange of a force particle and emerge from this interaction as 3 and 4.

But a very different process is also possible, this time represented by what physicists call *t*-channel scattering. In this case, as 1 and 2 approach each other they exchange a particle, which is an expression of the force or interaction between them. As a result of this change of particle, 1 and 2 modify their identities to become 3 and 4.

In other words *s*- and *t*-channel scattering have totally different physical interpretations, one in terms of the creation of a new resonance and its subsequent disintegration, the other in terms of an exchange of some elementary particle. Nevertheless, seen from a distance, all that happens is that 1 and 2 enter the interacting region, and 3 and 4 emerge. The two, *dual*, accounts of what happens inside are totally equivalent. Indeed it does not matter what happens inside the scattering region, provided one calculates the interaction using one or the other approach. This was a considerable puzzle for particle physicists; how can two very different physical processes end up producing identical results? And while Veneziano was unable to produce a deeper physical account for such results, at least he could write down a formula that reproduced all the results of this sort of scattering.

Veneziano's dual resonance model swept through physics. At last it seemed that order was being produced in the confusing world of hadrons. Admittedly each new scattering experiment produced a host of resonances, but now it was possible to relate these resonances to other quantum processes. I can remember attending a conference at Brandeis University in 1970 at which all the talk was about Regge poles and dual model resonances. Physicists felt that at last they were on the right track.

But despite the success of the dual resonance model, it was becoming clear that the theory contained a major defect—it allowed for the existence of ghosts. These ghosts are not the dead scientists of yesteryear, but what physicists call states of negative norm. They are a little like outcomes that have negative probability, which is clearly an absurdity. There may be a 30 percent chance of rain tomorrow. During a hot dry spell there may be only a 5 percent or even a 0 percent chance. But never a −50 percent chance; the idea of a negative probability does not make sense. Similarly the existence of ghost states was physical nonsense and a major flaw in the dual resonance model.

There turned out to be an ingenious way of getting rid of ghosts. This was to reformulate the theory, not in our usual space-time of four dimensions, but in a new space having twenty-six dimensions. (These twenty-six dimensions were also to play an important role in early string theory.)

But what is the meaning of a space with twenty-six dimensions? While it is true that physicists have the mathematical freedom to work in spaces of as many dimensions as they please, at some point they must return to reality and compare their descriptions with experimental observation. Attempts had been made several decades earlier to extend general relativity by adding an extra dimension and writing the theory in five instead of four dimensions. But it was a major leap of abstraction to add an extra twenty-two dimensions and reformulate a theory in a twenty-six-dimensional space!

In the end it became a choice between entertaining ghosts or working in a multidimensional space, and physicists, who are traditionally horrified of the metaphysical, chose the latter. Their expectation was that somehow those extra dimensions would eventually be explained away as a mathematical artifact. Physicists even attempted to produce an apology for the existence of these extra dimensions. Their idea was that these extra dimensions are not really spatial in the everyday meaning of the word; they are not the sort of dimensions through which material bodies can move on their orbits and paths. Rather they are "internal dimensions" belonging to the abstract spaces in which the symmetries of the elementary particles are defined. While a twenty-six dimensional space is needed for the dual resonance model, it could turn out that twenty-two of these dimensions will never be experienced by us in our large-scale world.

The idea of some sort of abstract space for the elementary particles had been known for a long time. A variety of patterns had been used to classify elementary particles into various families. It turned out that the relationships between various family members looked very like the sorts of relationships you get between parts of a spatially symmetric object, such as the arms of a starfish. However, since these relationships are not expressed in our everyday space-time, they are not really symmetry relationships in the normal sense of the word. But physicists seemed quite happy that these various elementary particle symmetries could be expressed at all. Early in the 1970s, physicists were asking if these abstract or internal spaces could somehow be related to the extra dimensions demanded by the dual resonance model.

The interpretation of extra spatial dimensions will come up again and again in this book. Many physicists are happy with the idea of multidimensional spaces, and today will accept the ten (no longer twenty-six) dimensions required of modern superstring theory. Others, like Roger Penrose, object strongly and feel that this is a "copout." It is sufficient to realize at this point that once the extra dimensions are permitted, then Veneziano's dual resonance model appears to work. It can relate information and data about scattering experiments to the observed resonances and in this way produce a coherence in the world of elementary particles.*

The dual resonance model also indicated that resonances and elementary particles behave as if they are extended in space. Would it be possible to relate this aspect of the model to the quark theory of elementary particles? Could it be shown, for example, that quarks come together so as to create extended objects that would also conform to the predictions of the dual resonance model?

It was at this point, in 1970, that Yoichiro Nambu came up with a striking new concept: Elementary particles are not points or quarks or even blobs, but vibrating, rotating strings.

In fact such ideas were already in the air at the time.

---

*More recently physicists have become aware of some powerful properties that apply only in a four-dimensional space. There may well be compelling reasons why a four-dimensional space-time is the true dimensionality of space. These topics are discussed at greater length in the final chapter.

Physicists had been hard at work trying to understand the meaning of duality. Why is it that what looks, from the perspective of the s-channel, like the disappearance of two particles and the creation of a resonance can look, from the *t*-channel, like a simple interaction?

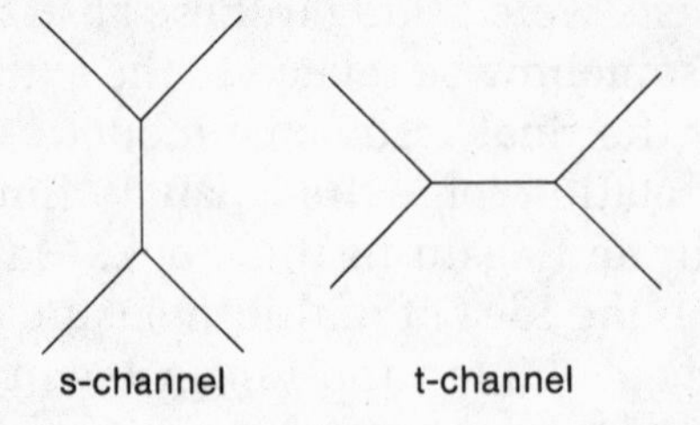

Figure 2–8
The interaction of two elementary particles can be viewed in two different ways. In *s*-channel scattering, an intermediate resonance state is created, while in *t*-channel scattering, a force particle is exchanged. In either case the same particles enter and leave the region of interaction.

But suppose that elementary particles are not really points but more extended regions. Why not smear out the preceding diagrams so they do not consist of lines but of sheets? The two scattering diagrams now become:

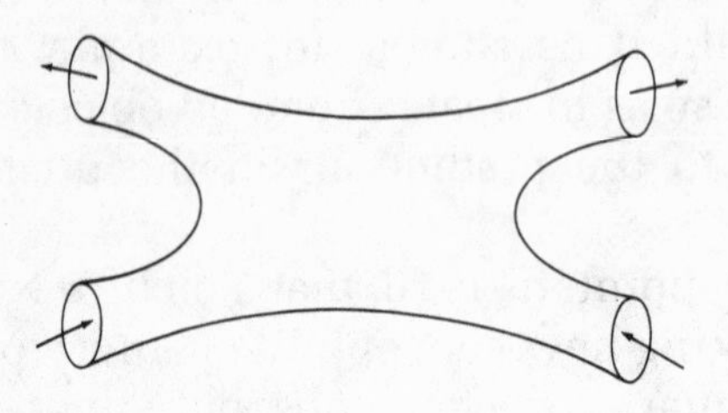

Figure 2–9
When elementary particles are no longer treated like dimensionless points, then the pictures of *s*- and *t*-channel scattering become topologically identical.

Since theoretical physicists had shown that the value of the interactions depends only on the topology of the sheets, then it immediately becomes clear that these two diagrams are totally equivalent. From the perspective of extended objects in space-time, the *s*- and *t*-channels are one and the same thing.

Already in 1969 some physicists were thinking in terms of the elementary particle not as a point but as a stringlike object. It was Nambu, however, who in 1970 pulled these various ideas together by expressing the meaning of a string in a clear, mathematical way.

Heavenly Harmony

Yoichiro Nambu was a distinguished physicist when he proposed the idea of string theory to explain the hadrons. As a young man he had carried out his first research at Tokyo University before going to the Princeton Institute of Advanced Studies and then to the University of Chicago. His investigations had taken him into such fields as the meaning of internal symmetries and how they are broken; he had introduced the idea of color* into the world of quarks; and, finally, in the 1970s he discovered a new way of describing the elementary particles. They are not point particles but vibrating, rotating strings.

But how can it be that the ultimate entities of our world are strings? In a sense, the story begins 2,000 years earlier with Pythagoras, who was responsible not only for a famous proof about the sides of right-angled triangles but for a philosophy that was centered around the power and harmony of numbers. The Pythagoreans taught that the mystical *tetrakyts*, or fourness, holds the secret to the universe. The *tetrakyts* can, for example, be used to construct a perfect triangle. Since 4 contains

*The term *color* should not be taken to mean that the elementary particles have real colors. Rather, color was a whimsical term adopted by physicists to describe certain characteristics of the quarks.

within it the numbers 1, 2, 3, and 4, so too the *tetrakyts* unfolds in the following way:

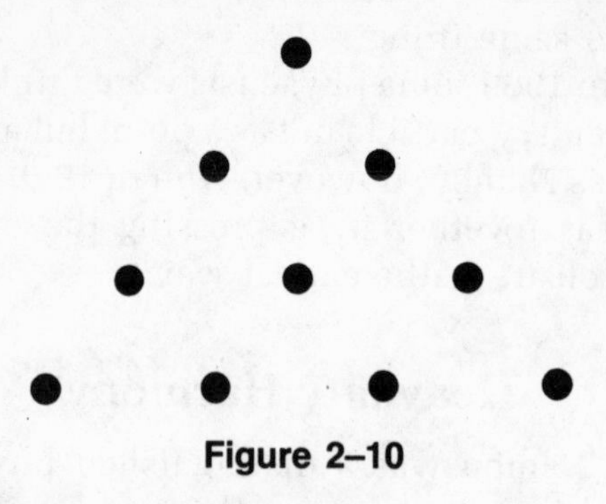

Figure 2–10

In following this philosophy, the Greeks sought to create all manner of ratios and harmonies of number. These harmonies, they believed, resonate throughout the universe. In particular, they are present in music.

Take a guitar or violin and place your finger midway along the string. The note produced is exactly an octave higher than the fundamental, played on the open string. With your finger two-thirds of the way along, the note is a fifth above the fundamental—a very pleasing interval. At three-quarters of the length, a fourth is sounded. Proceeding in this fashion Pythagoras discovered the diatonic scale, which is the basis of much of the world's music. Sound the notes of this scale together, and a pleasing harmony is produced. Move away from the fundamental on a journey through its various notes, and melody results. All, it appears, is generated by simple numerical ratios of length.

There is something magical about this result. Why should the beauty of number correspond with the melodies and harmonies that appeal to the ear? Why should there be such a deep link between music and number? In our own century, the theoretical physicist Eugene Wigner has drawn attention to what he calls the unreasonable effectiveness of mathematics. Again and again, abstract and beautiful mathematical relationships, explored for their own aesthetic sake, are later discovered to have exact correspondences with the real world—a coincidence that is quite remarkable.

Pythagoras believed that this numerical harmony extends throughout the universe. The distances of the planets, he argued, must also lie in simple numerical ratios. And, since the same ratios that govern the diatonic scale also determine the heavenly spheres, the whole universe moves with an abstract celestial music—"the music of the spheres." This harmony was confirmed when Kepler discovered his famous laws of planetary motion in which simple ratios hold between the length of each planet's year and its distance from the sun.

Since the Pythagoreans taught that the order of notes of a vibrating string pervades the whole universe, they would have been moved to learn, two thousand years later, of the work of Nambu (the theory of strings also contained contributions from Leonard Susskind at Yeshiva University and H. Neilson of the Neils Bohr Institute in Copenhagen), who was able to show that the hadrons, or strongly interacting particles, are also quantum manifestations of this Pythagorean ideal.

To return to the guitar string for a moment, the fundamental note involves a vibration of the whole string, with maximum displacement occurring midway along the string. Place the finger at this midpoint, and the vibration splits in two. Two vibrations must now fit into the space of one, and this is done by doubling the frequency—the number of times the string vibrates back and forth each second. The resulting sound is therefore an octave higher. When a finger is placed two-thirds of the way along the string, three vibrations have to fit into the string length.

A simple formula relates the frequency f of a note to the length L of the string, its mass m, and its tension T.

$$f = \frac{1}{2L}\sqrt{\frac{T}{m}}$$

Decrease the length, and the frequency increases. Twist

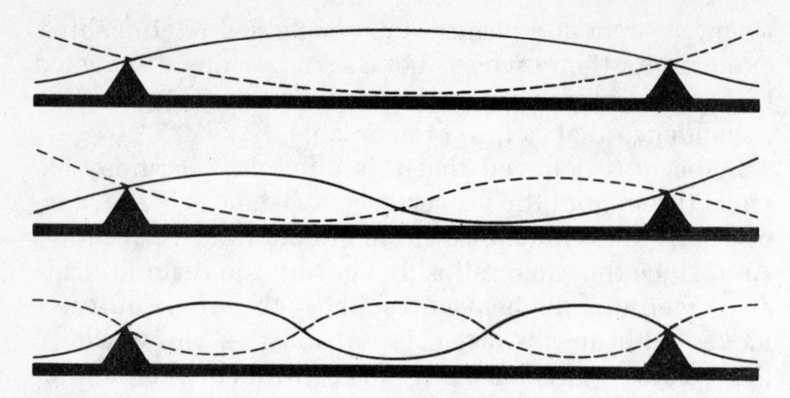

Figure 2–11
The scale of notes produced by a vibrating string corresponds to a series of standing waves.

the guitar peg to increase the tension, and the frequency of the note increases again. Substitute a heavier string, and the note drops.

Can this string produce notes that are not in tune with its harmonics and overtones? It turns out that such notes require vibrations that do not fit exactly into the string length. If such vibrations were ever produced, then they would rapidly die away.

The example of the guitar string suggests that an extended object can store energy in a series of discrete modes, like musical notes. It makes an interesting experiment to take a second guitar, in tune with the first, and pluck its string. Sure enough, the first guitar begins to sound in sympathy. It is possible in this way to transfer sound energy from one guitar to another, to feed the first guitar with enough energy that its strings begin to sound. Notes that take up energy in this way are said to resonate. But if the second guitar happens to be out of tune with the first, then a transfer of energy cannot take place. The strings of the first guitar do not resonate with those of the second.

The picture begins to have something in common with the dual resonance model. If the pattern of elementary particle resonances in a scattering experiment are indeed related to the vibrations of a quantum string, then we would certainly expect to see a simple pattern of resonances. Each resonance, or localization of energy, would correspond to a note of the quantum string. What appear to be elementary particles would in fact be the notes or the localized vibrations of quantum mechanical strings. (In fact, the strings proposed by Nambu have an additional freedom, for they can rotate as well as vibrate. But the restrictions imposed by quantum mechanics meant that these rotations *also* produce a fixed quantized pattern of "notes.")

Of course, there is a great difference between a metal guitar string and Nambu's strings that rotate and vibrate at the subatomic level. To begin with, Nambu's strings had to be very short; otherwise physicists would have detected their size in elementary particle–scattering experiments. Nambu decided to set the length of a string at around the experimental size of an elementary particle, some 10^{-13} cm. Unlike guitar strings, these quantum strings had to conform both to the constraints of general relativity and to the quantum theory.

But important similarities do remain. Going up a scale of notes on a guitar means that we are increasing the frequency of vibrations. In quantum terms, this increase in frequency means that the particles and resonances have more energy and therefore greater mass. An immediate implication of this simple picture is that the masses of the elementary particles and the energies of the resonances must conform to simple numerical patterns. In other words, Nambu's string theory accounted for the trajectories first discovered by Regge, those graphs of the positions of various resonances that were drawn to explain the results of high-energy scattering experiments.

In essence, Nambu had proposed that the origin of the hadron particles lies in tiny quantum strings that, in their rotations and vibrations, produce a pattern of quantum notes. These quantum notes, or localized energy levels, can be identified with the elementary particles and resonances. Just as musical notes lie on a scale, so too the particles and resonances are related. In a single step, Nambu had produced a highly intuitive picture that created a considerable degree of order in the world of elementary particles.

3
Nambu's String Theory

NAMBU'S STRING THEORY pointed the way toward a totally new conception of elementary particles as objects extended in space, whose internal excitations give rise to a series of energy levels and resonances. Unlike earlier theories, which pictured elementary particles as balls or diffuse blobs, these strings were one-dimensional objects, and being free to rotate and vibrate, they produced a rich spectrum of quantum notes.

But how were Nambu and the physicists who came after him to go from this initial intuition of a spinning, vibrating string only 10^{-13} cm long, to a full mathematical account of the hadrons, the heavy, strongly interacting elementary particles? The fine details of string theory will be left until Chapter 5, where they are explained in terms of the more successful superstring approach. But, in essence, Nambu first had to ensure that the string obeyed the constraints set by Einstein's general theory of relativity. In physicists' terms, the string equations had to be *manifestly covariant*. This means that the same laws of string physics must be seen by all observers, no matter how they are moving. In mathematical terms, the formalism has to be written down in such a way that it remains essentially unchanged when transformed from one observer's coordinate system to that of another.

When it comes to dealing with extended objects, general relativity can run into trouble, so care had to be taken to get the string formalism exactly right. In fact, it turned out that in order to satisfy the requirements of relativity, the ends of the string had to whip around at the speed of light. For this reason Nambu's string was also christened a "light string." And since material bodies cannot move at the speed of light, this means that the string must be massless. In a sense it becomes a generalization of a ray of light, a ray that can vibrate and spin.

String theory has to explain the dual resonance model, and this means the whole pattern of particle and resonance masses must be accounted for. But since Nambu's strings are massless, does this mean that relativity has thrown the baby out with the bathwater? In fact, the answer lies in the string's ability to rotate and vibrate. In singing, the higher you go up the scale, the more energy you need to produce the notes. In an analogous way, the quantum notes of the string—its quantized vibrations and rotations—are steps in a ladder of energy. But we know that energy has an associated mass, given by $E = mc^2$. So each excitation, or note of Nambu's string, has an associated mass. Although the light string has no mass itself, its quantum notes do.

With the string conforming to the restrictions of general relativity, the next step is to apply the rules of quantum theory. It turns out that a quantized string is not free to rotate in any way it likes, for its rotations can take place only in discrete, quantized units. This has the effect not only of associating each vibrational note with a quantum of energy, but of producing a pattern of notes for the rotations as well.

The vibrational quanta are widely spaced, like the rungs on a giant's ladder. Corresponding to each of these giant rungs we can erect a smaller ladder whose closely spaced rungs correspond to the string's rotations. The result is like having a series of musical scales that represent the vibrations and rotations of the strings.

Quantized light strings therefore produce a characteristic spectrum of energy and—since energy has an as-

Figure 3–1
Corresponding to each of the vibrational energy levels of a string is a finer-spaced series of rotational levels.

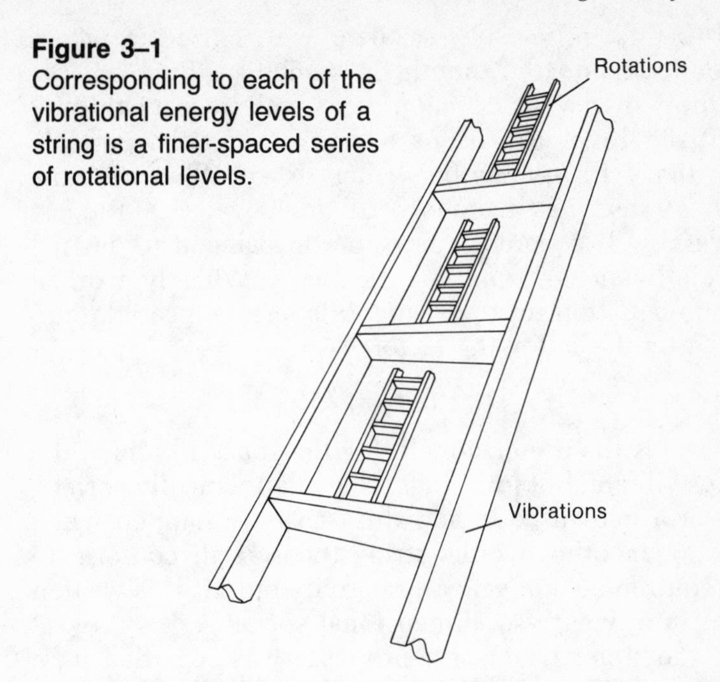

sociated mass—a distinct pattern of masses. It was the aim of string theory that these patterns of mass should correspond to the elementary particle masses and to the Regge trajectories that had originally led Veneziano to propose his dual resonance model. To achieve this correspondence, the length of the string was set at 10^{-13} cm, the limit on the size of the proton established in scattering experiments. Its tension was fixed at fifteen tons. With these two parameters established, the string produced a pattern of vibrations and rotations, and therefore a corresponding spectra of masses, which looked just right. Drawn on a graph, each spectrum of masses rises in exactly the same way as one of Regge's trajectories. String theory seemed to have captured the essence of Veneziano's dual resonance model. By assuming that at very small distances nature takes the form of massless, vibrating strings whose ends whip around at the speed of light it had been possible to create a picture that reproduced many of the characteristics of the known hadrons.

After this initial success, it now was time to take a closer look into the mouth of Nambu's gift horse. Demanding that a string should be both relativistic and quantized involves serious restrictions on the possible ways that a theory can be written down. What were the implications for string theory? Claude Lovelace at Rutgers University discovered a term that threatened to destroy the relativistic covariance of the theory. When he worked out the equation for the string, this term appeared, multiplied by the following factor:

$$[1 - (D - 2)/24]$$

where D is the dimensionality of the space in which the strings vibrate. But for things to be relativistically correct, this term had to be 0, and this can only happen when $D = 26$. In other words, string theory will conform to the requirements of general relativity only if it is written down in a twenty-six-dimensional space!

At the time, Lovelace's proposal was regarded as a crazy idea. What sense did it make to talk about physics in twenty-six dimensions when everyone knew that the world around us is four-dimensional? How on earth could scientists take seriously a theory that called for a one-dimensional object vibrating in a space of twenty-six dimensions? But when scientists were later able to show that those troublesome ghosts—paradoxical results predicted by the theory—that were haunting physics would also vanish in twenty-six dimensions, Lovelace's crazy idea began to take on a new light.

Suddenly physicists were willing to go along with twenty-six dimensions, provisionally at least. After all, they argued, what is so special about four dimensions? Our universe looks as if it contains three spatial and one time dimension. But is there any physical law that forces this to be so? Could it be that our universe is really twenty-six-dimensional but with twenty-two of these dimensions curled up so tight they can never be seen? If twenty-six dimensions are forced on us by nature then maybe there are ways in which twenty-two of these dimensions can be hidden. At least for the first time in science, nature seemed to be forcing a particular dimensionality, albeit a curiously high one, onto a physical theory. Roger Penrose, as we have seen, did not accept such arguments. If a theory can only be saved by dragging in twenty-two extra dimensions, he would probably say that the best thing is to throw away the theory!

Strings, Blobs, and Twistors

Nambu's approach certainly looked provocative. But why stop at strings? If one-dimensional objects work, then why not two- or three-dimensional ones? Why not vibrating discs, balls, or blobs? A number of earlier attempts had indeed been made to describe the elementary particles as three-dimensional, but they had always run into trouble. Serious problems arise with making three-dimensional objects properly relativistic and then quantizing them. In the end, string physicists believed that one-dimensional objects are the only possible attractive starting point.

It is one of those coincidences of physics that at the same time this string research was taking place, Roger Penrose was working at Birkbeck College in the University of London, developing his ideas of twistors. (Penrose later became the Rouse Ball Professor of Mathematics at Oxford University.) Twistors are also massless and one-dimensional; they too can be thought of as a sort of generalization of a ray of light. Yet in other respects, Penrose's approach was very different from that of Nambu. To begin with it was formulated not in a twenty-six-dimensional space but in four complex dimensions. And here the word *complexity* refers to complex numbers and not to difficulty. Remember that complex numbers are generalizations of ordinary real numbers. They contain both real and imaginary parts, where the imaginary parts refer to $\sqrt{-1}$.

At the time, there appeared to be little connection between twistors and strings. Indeed, while strings were essentially a topic in elementary particle physics, twistors were being studied by relativists, and at that time, there was little connection between the two groups. For two decades, these approaches were to develop in parallel; only relatively recently have physicists thought of putting them together.

Interacting Strings

A good theory of the elementary particles must also account for how they interact. Up to now physics has been forced to invent mechanisms for interactions or, in the quantum case, for an exchange of force particles, such as the photon. When it came to strings, things turned out to be far more straightforward.

What happens when two strings meet? How will they interact? Strings, physicists guessed, interact by splitting and joining their ends. Two strings join ends to form a single string. A single string splits in two. The idea is simplicity itself and requires no additional assumptions, no new forces or particles, simply the splitting and joining of strings.

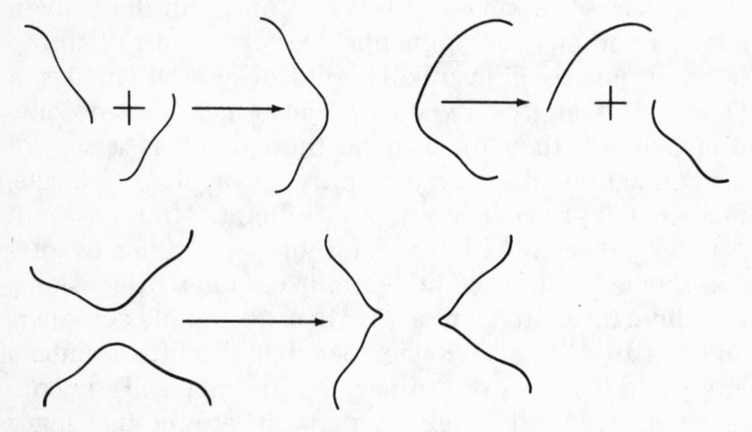

Figure 3–2
Strings interact by joining and splitting.

When physicists worked out the mathematics of string interactions they discovered that if the string is viewed from a great enough distance—and this turns out to be equivalent to viewing the string at low energies—it no longer looks like an extended object but appears to be a point. At this limit the splitting and joining strings look very like the exchange of force particles between two point particles—the strings. So it becomes possible to think of the traditional interactions between elementary particles as a sort of low-energy limit of the more interesting splitting and joining of strings. In addition, certain problems that had been associated with these forces were now found to clear up in twenty-six dimensions. This picture of the strong interactions, in terms of strings, was an exciting spin-off from the original theory.

If two strings can join ends, then why can't a string join its own ends to form a closed loop? There seemed to be no reason why this should not happen, so closed loops were added to open-string theory. But what did these closed loops represent? What sort of hadrons look like closed strings? It was at this point that the interpretation of string theory ran into difficulties. Physicists guessed that, like ants that meet on a trail, strings recognize each other by their ends. The general idea was that quantum numbers, which act as labels, are attached to the string ends. But what sort of labels can there be for a closed string? Having no ends, it would appear to have no identity and be nothing but empty space.

Further investigation confirmed that closed loops look like spin 2 vector bosons. And these very particles had earlier been hypothesized as the carriers of the gravitational force. Spin 2 vector bosons are the quantized units of gravity. And gravity is the curvature of space-time. Suddenly string theory had entered a very different field. It had begun as an attempt to explain the hadrons, the strongly interacting elementary particles, and their families, and now it seemed to be giving a picture of gravity as well. What on earth was gravity doing in a theory of the hadrons? The full answer remained hidden until Green and Schwarz's new superstring theory burst on the scene in the 1980s.

Strings and Infinities

String theory in twenty-six dimensions had many attractive features, but would it really avoid all those problems that had plagued conventional point particle theories since the earliest days of quantum field theory? Would the infinities, anomalies, and ghost states be forever banished from physics?

Physicists discovered that when string theory is written in a space of more than twenty-six dimensions, ghosts abound. Yet with fewer than twenty-six, ghosts come out of the woodwork and can only be removed by making the sorts of assumptions that do not sit well with the physics community. So ghost-free states demand twenty-six dimensions. Again that curious number 26 makes an appearance!

And what about those infinities that crop up when physicists attempt to sum up all the terms in a Feynman series—would they still be present in string theory? Many physicists believed that infinities are connected with the fact that certain terms are defined right down to an infinitely small region of space. But if physics is expressed in terms of strings, which have a finite length, then it is possible that these calculations could work out in a finite way. Unfortunately infinities continued to plague the new string theory—indeed it was only with Green and Schwarz's breakthrough into the world of superstrings that these infinities were finally abolished.

String theory also had the unfortunate characteristic of predicting the existence of elementary particles that move faster than light. It turns out that such particles are not in fact forbidden by the theory of relativity. Einstein showed that the speed of light acts as a barrier no material body can transcend. In fact, only the massless photon and neutrino can move at the speed of light. But there is nothing in Einstein's theory to prevent bizarre objects from already existing on the other side of the light barrier. These hypothetical tachyons, as they are called, cannot have normal masses; their masses must be imaginary (in the mathematical sense of the word).

Tachyons can be thought of as particles in a mirror world, reflected by the speed of light. Just as the speed of light is a barrier to particles in our universe (matter can never be accelerated to reach this speed), it is equally a barrier to the hypothetical tachyons. Tachyons could never slow down enough to reach the unattainable speed of light.

Although tachyons are not absolutely forbidden by relativity, they are something physicists would rather do without. If tachyons really exist, then they would play havoc with our idea of causality. Physicists therefore ruled out their existence and then tried various ways of getting rid of the tachyons from string theory. But it was only with Green and Schwarz's superstring approach that they were finally expelled.

Strings and Quarks

Of course, string theory was not the only game in town. In 1964 Murray Gell-Mann at the California Institute of Technology and, independently, George Zweig in Geneva had proposed that the elementary particles are built out of more primitive entities called quarks. On the other hand Nambu was suggesting that hadrons are vibrating, rotating strings. Was there some way of connecting the quark and the string pictures?

One idea was to smear the quarks over the whole string—possibly the quark *is* a string. But this approach ran into trouble when very high-energy scattering experiments indicated that the hadron had an internal "gritty" nature. The hadron behaves as if it contains tiny grainy particles, and physicists guessed that these must be the quarks, so smearing quarks over much longer string distances did not make much sense.

Another approach was that the ends of a string are quarks. This was a neat idea, for it also explained why quarks had never been seen. After the birth of the quark theory, many experimentalists had attempted, without success, to detect free quarks. In the end physicists were forced to conclude that quarks can never be seen. But why should this be?

String theory provided a simple answer. Quarks are the ends of strings. Break apart a string, and what do you get? New ends. It is no more possible to have a free quark than it is to have a string with only one end.

But when the details of this idea were worked out, things did not look so attractive. A meson is made of two quarks—a string with two ends. But what about a proton? It is supposed to contain three quarks; how can a string have three ends? Moreover string ends whip around at the speed of light. Physicists did not much like the idea of a charged particle like a quark moving at the speed of light.

Bosonic and Fermionic Strings

There was yet another problem with string theory. It was designed to describe the hadrons, and hadrons are all fermion particles, having fractional spins like $\frac{1}{2}$ or $\frac{3}{2}$. But when physicists took a close look at Nambu's strings, they discovered that the strings had integral spins; they were bosonic objects.

This was an embarrassing state of affairs. While the strings appeared to duplicate essential features of the mass spectrum of the hadrons, they belonged to entirely the wrong family! Was there some way of creating fermionic as well as bosonic strings?

In 1971 Pierre Ramond at the University of Florida discovered a way of getting over the problem. Ramond showed that it was possible to write equations for strings with fractional spins. He did this by making use of what are called spinors. These spinors are the same mathematical objects that are used to describe the spin of the electron, proton, neutron, and other fermion particles. If spinors are sufficient to describe the fractional spins of point particles, then why not adapt them to string theory and use them to describe fermion strings?

But Ramond also knew that his strings were capable of joining up, and when two fermion strings with half-integral spins join together, they must generate something with an integer spin, namely a boson. In other words, a theory that attempts to describe strings as fermions must also be prepared for the sudden appearance of bosons as well! But could bosons be incorporated mathematically into Ramond's string theory?

John Schwarz, at that time a postdoctoral student at Princeton, and André Neveu, in France, attempted a formulation that would be similar to Ramond's but this time would explicitly include bosons. The result was picturesquely named a "spinning string theory." (Contributions were also made by Joel Scherk, a French physicist working at the California Institute of Technology. Scherk was one of the founding fathers of modern superstring theory, but he died too young to see his ideas fully blossom.) The new spinning string theory, which could take account of hadrons as fermions, did not require a twenty-six-dimensional space for its formulation but could be written down in ten dimensions.

The fact that fermions and bosons could both be thought of in terms of the vibrations of spinning strings caused some physicists to wonder whether there could be a deep connection between these two classes of particles. Traditionally fermions and bosons had been kept firmly apart as the two great families of elementary particles—rather like the kingdoms of plants and animals of the living world. But now physicists were also beginning to wonder if it would be possible to link bosons and fermions together with a new symmetry called supersymmetry.

Finally, in 1976, Joel Scherk, along with F. Gliozzi and D. Olive, showed how fermions and bosons could emerge on an equal footing out of the one spinning string theory. In their approach the various states of the string could be thought of as bosons or fermions, with every boson having a fermion for a partner. Fermions and bosons were paired up at last, the two great families of the elementary particles became reflections of each other. Strings had become supersymmetric; there were, in a sense, superstrings!

If physics had continued to push ahead with this approach, it is possible that a true superstring theory could have been created in the 1970s. But physicists at the time were becoming far more interested in some of the new ideas that will be discussed in the next chapter, ideas like supergravity, for example. In Michael Green's opinion, this 1976 paper of Gliozzi, Scherk, and Oliver was the "dying gasp" of the first era of string theory. In fact, the authors had only used the string formalism as a way of working toward some new results in supergravity theory. After 1976 strings were more or less forgotten as physicists moved into newer and what appeared to be more exciting areas.

Conclusions

While Nambu's string theory had begun directly enough, as a theory about the nature of hadrons, it had now developed in some unexpected ways. Not only did it describe hadrons, but it seemed to be saying something about gravity as well. It produced a spectrum of particles that looked like the hadrons, but in some formulations, it was also plagued with ghost states, tachyons, and infinities. It accounted for the interactions of the elementary particles in a new and novel way. But it also demanded twenty-six dimensions (or was it ten?) for its formulation. As one physicist put it, string theory began by trying to explain the hadrons and ended up accounting for only two particles—the pi and the rho mesons. But the pi meson turned out to be moving faster than light, and the rho meson had no mass!

Physicists could, of course, have pushed on with the string theory, trying to clear up these problems. But already important events were taking place in other areas of particle physics. The weak and electromagnetic interactions had become unified into a single electroweak force, and there was an expectation that this force could be further unified with the strong nuclear force. Scherk's supersymmetry approach had been revived (while at the same time throwing away the underlying strings) and was providing an additional way of unifying the elementary particles by linking fermion to boson. Attention was turning back to the point particle theories in their new grand unified, supersymmetric, supergravitational forms. Suddenly strings were old hat, and no one really wanted to know about them. No one, that is, except for people like John Schwarz and Michael Green, who continued to worry about Nambu's ideas and where they could have gone wrong.

4
Grand Unification

Unity in Physics

THROUGHOUT ITS HISTORY—which is a relatively short one when compared with other great human endeavors such as art, music, drama, poetry, and philosophy—science has pursued a Holy Grail called unification. The truly great scientific minds have always been concerned with discovering a unifying pattern to phenomena and bringing ideas together within the compass of a single insight.

Physics in the eighteenth century, for example, involved a number of separate lines of inquiry such as mechanics, magnetism, the phenomenon of static electricity, and the nature of light. Natural philosophers were able to work in one area without having to bother about the others. But during the nineteenth century, the concept of energy was developed to the point where it suggested a way of unifying several of these branches. Scientists realized that energy is involved in a burning fire, when water turns into steam, when steam drives an engine, when that engine turns a generator to make electricity, and when electricity produces light. Although the studies of mechanics, chemistry, light, electricity, and various changes of state were initially quite separate, there had now appeared a unifying principle that was capable of connecting them together. The single concept of energy is involved when matter changes state from solid to liquid to gas; it is also released in chemical reactions like combustion; it is responsible for the operation of machines and the production of electricity and light. In this way the unifying science called thermodynamics enabled scientists to interconnect different phenomena under the one umbrella of the transformations produced by energy. In 1905 Einstein presented a further unification when he showed that even energy is not separate from matter, since the one can be transformed into the other and back again.

Another masterly stroke of unification came with the realization that magnetism and electricity arise as aspects of one underlying phenomenon: electromagnetism. For hundreds of years phenomena such as lightning, static electricity, and magnetic attraction had been observed and studied. But it was only during the eighteenth and nineteenth centuries that scientists began to realize that these two domains of knowledge—electricity and magnetism—had something in common. An electric current will deflect a magnetic compass needle; likewise, when a magnet is moved past a coil of wire it produces an electric current. In other words, electricity affects magnetism, and a moving magnet creates electricity. It was the Scottish physicist James Clerk Maxwell who was finally able to bring together these two phenomena within a single theory called electromagnetism. In fact, his theory produced, gratuitously, an additional act of unification, for light itself was now treated as an aspect of an undulating, oscillating electromagnetic field.

Maxwell's unification was one of the triumphs of nineteenth-century theoretical physics. But it was left to Einstein to go even farther. Maxwell's theory was distinct from that of Newton's, for electromagnetism did not unite with the mechanics of moving bodies. One dealt with charges, fields, and currents, and the other with the movement of matter. Einstein realized that there were certain symmetries inherent in the mathematics of these two theories that were essentially incompatible. Yet because electrical phenomena have mechanical consequences, and vice versa, as the electrical motor and generator demonstrate, in some essential way the two theories had to unite.

It was within his special theory of relativity that Einstein was able to rewrite Newtonian mechanics in such a way that the symmetries and mathematical transformations within Maxwell's theory and the new relativistic equations of motion became identical. As a matter of fact, this is the very reason why the speed of light now enters into a theory about moving matter. Electromagnetism and mechanics become unified within special relativity.

The goal of physics has always been to unify in this way, to bring apparently dissimilar things together. But could the spirit of unification be seen in the particle physics of the late 1950s? Far from it. Physics at that time consisted of a large collection of theories, facts, speculations, observations, hypotheses, and data, none of which really fitted together in any truly satisfying and coherent way.

To begin with, there were too many particles which scientists were trying to fit into a number of families. Then there were the various rationales for the family groupings themselves, which were supposed to explain why certain particles turn up with given partners, or why some particles have very different masses from others. And what about the forces? Physics demanded four of them, but how are they related to each other and why do they have such different strengths? As we learned in the previous chapter, electromagnetism does seem to be internally consistent, but when it came to the strong interaction between hadrons, the force was far too strong to be calculated.

The electromagnetic interaction could be understood in terms of an exchange of virtual photons (which are sort of temporary particles, particles taken on loan, as it were): An electron borrows a little temporary energy from the quantum vacuum and then uses it to create a virtual photon, which shoots across to another electron nearby. The photon is then annihilated and its borrowed energy given back to the quantum vacuum. The two electrons therefore interact by swapping virtual photons. In a similar way the strong interaction could be understood in terms of an exchange of mesons. But what about the weak interaction? In the 1960s it looked as if this interaction did not involve any exchange of particles at all. Did this mean that the weak interaction is somehow different from that of the strong and the electromagnetic, with no possibility of unification?

In fact it later turned out that the weak interaction does involve the exchange of a particle, called the intermediate vector boson. But because the mass of this particle is so large and its range so short, it would have been impossible to see it at that time.

Scientists welcomed the dual resonance model and, following it, the early string theory, for these approaches promised to bring about a form of unification. Since many physicists believed that the elementary particles are the basis of all physics and, in turn, that physics itself provides the foundation for chemistry and biology, this unification was especially welcome.

But, in the end, even Nambu's string theory failed. It did have some attractive aspects, like picturing elementary particles as extended objects and providing a spectrum of particles and resonances from a single string. But, as it turned out, the actual particles it predicted were not observed in nature. Moreover it was plagued with objects such as tachyons, faster-than-light particles, and by those curious extra dimensions that some physicists found to be unphysical.

Therefore, despite their initial excitement, physicists became disenchanted with string theory. Already new ideas were on the horizon, such as the standard model, which had made great progress with an earlier idea that all hadrons are made out of more elementary particles called quarks. This was soon followed by the spectacular unification of electromagnetism with the weak interaction and then with the strong nuclear force. Unification of physics, it appeared, did not require strings. Physicists could manage quite well without the need for these hypothetical spinning, twisting, vibrating objects. The 1970s were about to dawn and with them the Age of Grand Unification. Strings looked as if they were truly dead and buried.

Quarks and the Standard Model

Right from the earliest days of quantum theory, physicists had tried to discover ways of relating the elementary particles to each other. The brilliant English physicist P. A. M. Dirac had, for example, hypothesized on purely theoretical grounds that every particle has its corresponding antiparticle with mirror-image properties. Dirac had to wait several years until an experimental physicist, Carl Anderson, observed curious electrons that appeared to have the wrong charge—positive instead of negative. These were the positrons, antiparticles to the electron. Some years later, the antiparticle to the proton—called the antiproton—was also observed. Today we know that every particle has its antiparticle. Particle and antiparticle is a fundamental symmetry of physics.

Another symmetry was discovered when quantum theory was still in its infancy. Even with the advent of this new theory, there were still problems in accounting for lines in the spectrum of elements such as hydrogen, lithium, potassium, and sodium. Close examination showed that each line was in fact made up of two closely spaced partners—there were twice as many lines as the theory predicted. In the end it was George Uhlenbeck and Samuel Goudsmit, both from the University of Leiden, who suggested that the electron must have a basic "two-valuedness" with the immediate consequence that it exhibits twice as many energy states as expected.

The next step in the puzzle was solved when the deeper reason for this two-valuedness was explained. Scientists realized that the electron could be thought of as having a quantized spin. This spin has only one of two possible directions—up or down—hence its basic two-valuedness. Corresponding to an electron with spin up, there can also be a state in which the electron is spinning in the opposite sense, spin down. The mathematical formulation of electron spin relied upon a quantum object called a spinor—the very same spinor that would later be used to create *spinning strings* and, by Roger Penrose, to produce giant spin networks and finally to be generalized into the *twistor*. Spin was another fundamental symmetry of the elementary particles.

In 1932, a few years after Uhlenbeck and Goudsmit's discovery, Werner Heisenberg began to think about the nature of the neutron and proton. At first sight these two particles look quite different; the proton has a positive charge (its antiparticle has a negative charge), while the neutron has no charge. In addition, the neutron is 0.1 percent heavier. But, Heisenberg speculated, what would happen if the electromagnetic field did not exist? In such a hypothetical world, there would be no way of experiencing the charge on the proton—it would look just like a neutron. And what about that slight difference in mass? Well, that could always be due to some sort of electromagnetic interaction.

Heisenberg guessed that if the electromagnetic field could be switched off, then the proton could not be distinguished from the neutron. In other words, since there *are* neutrons and protons, they could be thought of as a single particle having a basic two-valuedness. Suppose, therefore, that the proton, and the neutron possess something analogous to quantized spin, with two possible values. Spin up would signify a proton and spin down a neutron. With the electromagnetic field switched off, there would be a mirror symmetry between an identical particle with two opposite spins. Switching the field back on breaks the symmetry and picks out the proton from the neutron.

This new form of spin was called *isospin*. Since one form of spin already exists in our space, scientists hypothesized that isospin must exist in a new abstract space called isospace. Now particles could be characterized according to their electrical charge, spin, and isospin.

At first these three labels produced order within the multiplicity of the elementary particles. But soon additional labels were needed. Partners to the proton, called hyperons, were discovered, and a new label called "strangeness" was invented. (The term *strangeness* does not mean that elementary particles are in some way weird, rather it is one of those idiosyncratic choices of terms characteristic of twentieth-century physics. Like color, strangeness is a new way of characterizing the elementary particles.) Just as it was possible to place a mirror in isospin space and reflect the neutron into the proton, so too it was possible to erect a mirror in strangeness space and reflect the proton into the hyperon. If the two mirrors were held at right angles, then a multiplicity of reflections was produced. Physics was on the road to unification, and the quark model was about to be born.

But before going on to look at this, it is necessary to understand isospace, or strangeness space. What exactly is it, and what relation does it have to the space we live in? It is difficult to give satisfactory answers to these questions, for physicists until recently were by no means agreed on their meanings. Possibly these *internal spaces*, as they are called, are a pure mathematical device with no physical correspondence. On the other hand, internal space may be connected in some way to the space we live in. It was with the advent of superstring theory that physicists first learned how to put together internal spaces and space-time. Yet in essential ways, some of the mystery still remains.

And now back to those symmetry mirrors. By the early

1960s, Murray Gell-Mann and Yuval Ne'eman had discovered that it was possible to make patterns of the elementary particles in which each one could be obtained by a reflection in the strangeness or isospin mirrors. Reflecting objects in pairs of mirrors produces a particular pattern. The various relationships between the elementary particles in this pattern have the appearance of symmetry relationships. In fact, when the various symmetry relationships are collected together, they form what mathematicians call a symmetry group—in the case of reflections in the strangeness and isospin mirrors, this group is called SU(3).

The mathematics of this SU(3) group indicates that it can contain only patterns formed by 3, 8, or 10 particles. In fact, the octet or eightfold pattern turned out to be a very good way of grouping elementary particles together so that they all had roughly the same mass. This led to the name *eightfold way* as a description of this symmetry approach. (The term *eightfold way* also hints at the essential path of Buddhism.) It was also possible to group the mesons into an octet.

Then a series of resonances were discovered. There were nine of these, which suggested either that there was one too many to form an octet or that the resonances make a tenfold pattern that has one missing entry, a particle with definite properties that had not yet been observed. Such a prediction of a missing particle is a little like what happened when in 1871 Dmitry Mendeleyev arranged the chemical elements into a pattern. Mendeleyev's scheme also contained gaps, missing elements that had not been found in nature. It was the triumph of Mendeleyev's *periodic table,* as it became known, that these elements were discovered toward the turn of the century.

In a similar way, the patterns of the elementary particles received a great boost when, in 1964, the omega minus particle was discovered. It fitted right into the gap left in the pattern of resonances.

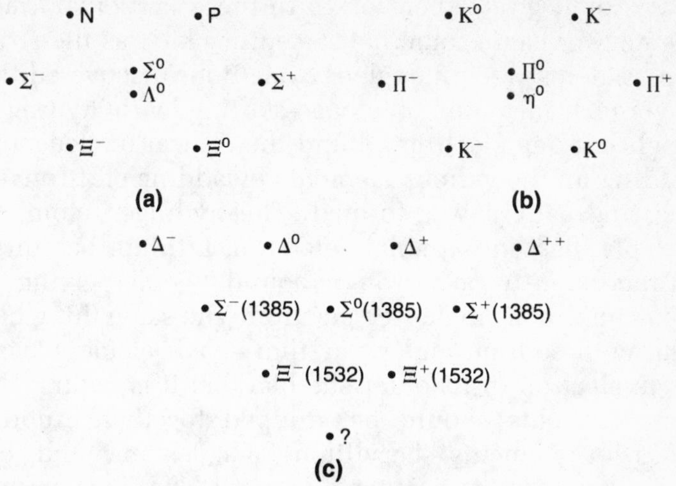

• N • P

• Σ^- • Σ^0 • Λ^0 • Σ^+

• Ξ^- • Ξ^0

(a)

• K^0 • K^-

• Π^- • Π^0 • η^0 • Π^+

• K^- • K^0

(b)

• Δ^- • Δ^0 • Δ^+ • Δ^{++}

• Σ^-(1385) • Σ^0(1385) • Σ^+(1385)

• Ξ^-(1532) • Ξ^+(1532)

• ?

(c)

Figure 4–1

(a) The neutron and proton are members of an eightfold family of heavy fermions.
(b) Eight mesons combine to form another octet.
(c) Nine elementary particle resonances form a pattern that suggests a missing member.

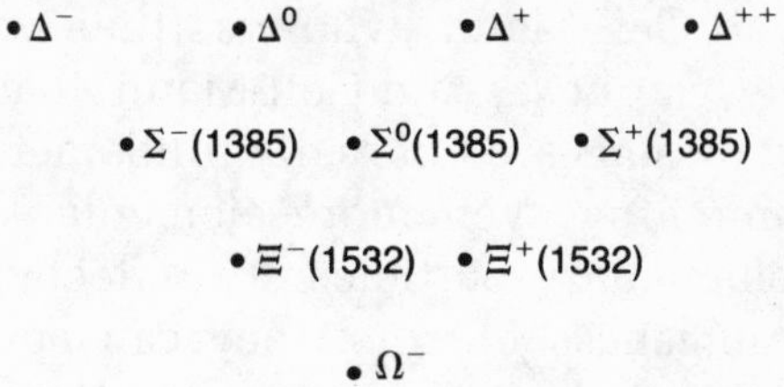

• Δ^- • Δ^0 • Δ^+ • Δ^{++}

• Σ^-(1385) • Σ^0(1385) • Σ^+(1385)

• Ξ^-(1532) • Ξ^+(1532)

• Ω^-

Figure 4–2

The omega minus particle fits into the gap of the ten-member pattern of elementary particle resonances. Here the constituents of these resonances are expressed in terms of up (*u*), down (*d*), and strange (*s*) quarks.

A further analogy can be drawn between Gell-Mann and Ne'eman's patterns and those of Mendeleyev. For while the Russian chemist had been able to relate the chemical elements in families according to their various properties, there was still no underlying reason as to why they should group themselves in these particular ways. The answer came some half a century later as the structure of atoms was untangled. Niels Bohr discovered that in going from atom to atom—starting with hydrogen, which has one electron around the central nucleus, and building up the various elements by adding electrons—a repeating pattern was formed. The hydrogen atom, for example, has only a single electron. Lithium has three, but the inner two are strongly bound together, giving the appearance of a single free electron. The same thing happens with sodium and potassium—each element has a single electron in its outer shell so that it is natural that these elements should be grouped together. Another group of elements—beryllium, magnesium, and calcium—have two free outer electrons. The inert gases—helium, neon, argon, krypton, and xenon—have no free electrons because their outer shells are filled.

So this understanding that chemical elements are themselves made out of something more fundamental (the atoms) produced an underlying reason for the patterns that Mendeleyev had observed. A similar insight occurred in 1964 to Gell-Mann, who was then working with George Zweig. Suppose, said Gell-Mann, that the proton, neutron, and mesons are not really fundamental but composed of something even more elementary. Remember that when elementary particles are reflected in the isospin and strangeness mirrors, they can be grouped together according to the symmetry SU(3). But SU(3) allows only for patterns containing three, eight, and ten particles. Physicists had already discovered eightfold and tenfold patterns—but what about a trio of particles? Gell-Mann speculated that this trio pattern in fact exists, and that it is even more fundamental than the other patterns of elementary particles. In fact, the threefold pattern represents even more elementary entities called quarks. Combining these three quarks, it would then be possible to reproduce all the properties of the eightfold and tenfold patterns.

The three quarks were labeled up, down, and strange. It was a simple matter to build a proton out of two up and one down quarks, while a neutron was composed of two down and one up quarks. In this way the hadron patterns could be produced and the properties of the missing omega minus predicted exactly.

The quark model turned out to be a triumph for elementary particle physics. It brought order into the world of the hadrons. It also became possible to think about the interactions between hadrons—called the strong interaction—purely in terms of interactions between quarks. It would also turn out that, rather than the quark-quark interactions blowing up, they would be quite manageable. In fact quark interactions were as well-behaved as electrodynamics had been. This would prove to be one of the great breakthroughs in dealing with the hadrons.

But the quark interaction was still down the theoretical road. A more immediate step was to produce quarks in high-energy accelerators and confirm their properties. The hunt for the quark was on.

A few years later, despite a wide variety of experiments which included looking for quarks everywhere from inside oysters to at the bottom of coal mines, physicists were willing to admit that quarks were totally elusive. Possibly there was some law of nature that prevented free quarks from ever appearing. String theory, in fact, provided one such explanation: Quarks are the ends of strings. So when a string is broken by a high-energy collision, it is no more possible to have a free quark than it is to have a free end to a piece of string. Broken strings always contain two ends—quarks bound together. But in the last analysis, the quark theory did not really have to rely upon strings to account for the elusiveness of quarks; physicists were able to produce a variety of alternative theories to explain the quark's absence from experimentalist's detectors.

If free quarks were never seen, physicists were willing to live with the fact. Moreover, by the 1970s very high-energy collisions of protons suggested that the hadron had a sort of internal "graininess"—tiny internal gritty pieces. Could these be the elusive quarks?*

Of course, quark theory did not stand still. Bigger and better particle accelerators produced ever more elementary particles. As a result, bigger patterns were demanded, and these required more reflections, additional symmetry mirrors. By the midsixties a new symmetry called color had been introduced, and six quarks had replaced the initial three.

Using the six-quark theory, the individual quarks were pictured as interacting via a new force of nature—the glue force. Or to put it another way, quarks exchange gluon particles between each other in order to interact. Just as electrons interact by the exchange of photons, so too quarks interact through the exchange of gluons. By analogy to the successful theory of quantum electrodynamics, this new theory was called quantum chromodynamics. At last the insoluble strong interaction between hadrons could be replaced by a well-behaved glue interaction between quarks.

Quarks and Leptons

If the elementary particles are to be fully unified, then what about the leptons? After all, the quark picture only accounts for the much heavier hadrons and leaves out

---

*An understanding of the forces that operated between quarks also provided an explanation for why no free quarks are ever seen. The glue force, as it is called, is unlike any other force in elementary particle physics. As with an elastic band, it gets stronger the farther apart the quarks are pulled. Try to pull two quarks apart to produce a free quark, and this force will become stronger and stronger. In fact, the energy needed to pull apart two quarks will be sufficient to create two new partners—hence no free quarks. On the other hand, when quarks are close together, as happens inside the proton, they interact so weakly as to appear free. Physicists called this condition asymptotic freedom. So when very high-energy experiments probe the interior of the proton, they see what appear to be free quarks. Hence the grainy appearance of the proton.

the lighter particles like the electron, muon, tau, and their neutrinos. High-energy experiments, however, indicated that, unlike the proton and other hadrons, the electron does not have an internal "grainy" structure. Possibly it really is elementary and not made out of any other particles.

Physicists discovered that the six leptons—electron, muon, tau, and their companion neutrinos—were actually the lighter partners of the quarks. Although this proposal raised a number of unanswered questions, it did suggest the possibility of a grand unification in which the entire material universe emerges out of a small number of elementary building blocks.

But unification was only partial, for there still existed four quite separate forces in nature: elementary particle physics required the existence of the glue, electromagnetic, and weak forces. A truly unified theory would deal with only a single force, and moreover, that force should finally be unified with gravity. The story of unification of the forces of nature begins with what is called gauge invariance.

Gauge Theory and the Forces of Nature

When I was a young boy, my father gave me a compass and told me that I could use it to find the direction of the North Pole. Now, while I could see that the compass kept its same orientation as I moved it around, I still couldn't work out which direction was north. The problem was that the needle was symmetric; it had a point at both ends. Admittedly one half was painted black and the other white. But which was which? Did the black end point north or south? This gift seemed to me to have a basic ambiguity or two-valuedness.

But suppose my friend across the street had also received a compass. He too would be in the same quandary of working out which point indicated north and which indicated south. Each of us would be faced with two choices. However, once we began to compare notes then we would have to be consistent in our choice. We could not both be facing in the same direction while one called it north and the other south. So by communicating with each other we would fix our convention.

A few decades later, physicists also began to worry about ambiguities concerning the direction of forces and fields in space. Their solution—which also applies to the compass needle—was called *gauge theory*. It was to become one of the most important insights within particle physics and the key to unifying the forces of nature.

What happens in the case of the compass needle is that there is a symmetry between north and south. Each person is totally free to choose which end of the needle to call north. Physicists call this type of symmetry, in which there is a freedom of choice at each point in space, a *local symmetry*. When, however, the symmetry choice at one point is identical to that at every other point, a *global symmetry* exists. Clearly a local symmetry offers more freedom of choice than a global symmetry.

With a local symmetry, like the compass needle, it only makes physical sense if scientists at different parts of space agree on a single convention. But this implies that nature must also send information from point to point, that a field of information can be transmitted across space. Scientists call such a field, which carries, for example, the convention about north and south, a *gauge field*.

Think of physicists at work on a spaceship trying to discover the basic laws of electricity. Scientists in some other ship also work on the same problem. Nature's consistency demands that both groups of scientists should come up with exactly the same laws. But suppose that one spaceship is charged up to a high voltage or immersed in a strong magnetic field—or even that all the positive charges on one ship are called negative. Since the laws of nature have an inherent local symmetry hidden within them, this means that scientists will always come up with exactly the same laws, even if the ships are at different electrical potentials and no matter what conventions they may choose locally.

But how is information sent from one spaceship to the next? In what way does nature itself keep track of its own local conventions? How does nature know that a positive or a negative charge is at some distant location in space? What is the *gauge field* corresponding to electricity and magnetism?

The answer is fascinating. The electromagnetic field is itself the gauge field for electricity and magnetism. Or to put it another way, when it comes to electromagnetism, nature, on the surface, *appears* to be globally symmetric. Electrons on earth are negatively charged just as they are on the moon. But within this global symmetry is hidden a more general local one. It is the combination of this local symmetry and the electromagnetic gauge field that produces the apparent global symmetry of nature out of the more general local symmetry. For if we switch off this electromagnetic field, then local ambiguity or two-valued choice is restored, and we have no way of knowing, for example, the sign of the charge on an electron located some distance away.

Physicists began to think about electromagnetism as a gauge field. Whenever nature exhibits some form of global symmetry—a convention about the nature of charges or quantum numbers that extends across space—it is probably the result of a powerful local symmetry plus a gauge field. Since these gauge fields must stretch across all space, in other words, have an infinite range, the fundamental particles associated with them must be massless. In the world of particle physics, the heavier the particle that is associated with a force field (or, in other words, carries the force), the shorter is the range of this force. Mesons, being quite heavy, restrict the strong nuclear force to around the dimensions of a nucleus. The electromagnetic field, by contrast, is carried by massless photons and therefore has an infinite range.

In looking for gauge fields, physicists had been looking

for massless boson particles (in fact, what are called massless vector bosons)—the quantum particles corresponding to the gauge field, which are also the carriers of the force. But so far, only one such massless boson was known: the photon, which is the carrier of the electromagnetic interaction.

Gauge fields are related to the structure of space-time itself. They are an essential arm of the whole principle of symmetry. What was previously taken as a fundamental force of nature—electromagnetism—is now seen as a gauge field, an implication of the local symmetry of the elementary particles. This is an exhilarating new vision about the nature of electricity, magnetism, and light. But could it be that what is true for electromagnetism is also true for the other forces of nature—for the strong and weak nuclear force, and even for the gravitational force? Are local symmetries the underlying reason for all the forces within the universe?

Electromagnetism can be understood in terms of what physicists call a gauge field. The next logical step, therefore, was to see if the strong force between quarks and the weak nuclear interaction could also be described in terms of gauge fields. Physicists Steven Weinberg, J. C. Ward, and Abdus Salam began to think about the weak interaction in an attempt to work out how it would be possible to picture it as a gauge field. If things had worked out in a conventional way, then the researchers would simply have come up with a new gauge field that would describe how elementary particles interact via the weak force. But, in fact, something very exciting occurred. To make their approach work, the physicists discovered that they could not use a single weak interaction gauge field alone, but needed a sort of unified field which describes not simply the weak interaction but the weak and electromagnetic interactions combined. This, however means that two of the fundamental forces of nature—electromagnetism and the weak nuclear force—must now be treated in a unified way. This new gauge field had to be described by a symmetry group called SU(2) x U(1) and is built out of the symmetry group SU(2), which describes the weak nuclear force, and U(1) for the electromagnetic field.

Physics had hitherto viewed these forces as quite distinct—after all, the weak nuclear force has to do with elementary particles, while electromagnetism is an infinitely ranged force at our own scale of things—but now nature was forcing physicists to treat these forces as one and the same. The result was a single gauge field to describe a new unified electroweak force. It turned out that this field has four components, and therefore four massless vector bosons, which carry the force. Two of these gauge particles have electrical charges, two do not. One of the massless vector bosons looks very like our old friend the photon.

What was the meaning of Ward and Weinberg's result? Nature, they seemed to be saying, knows only a single gauge field which embraces both the weak and electromagnetic interactions. But a host of experiments show very clearly that the weak interaction is different from the electromagnetic; it has a different strength and, far from being of infinite range, acts only over very short distances. And what about these massless charged vector bosons flying about at the speed of light, and the massless partner to the photon? At first sight, the cost of unifying two of the forces seemed too high. Putting the weak force on an equal footing with the electromagnetic introduced too much symmetry into the theory. Nature simply does not look that way.

It was at this point that another powerful idea was applied, that of symmetry breaking. In essence, this suggests that while the laws of nature have a high degree of symmetry, their particular solutions—in other words, the objects we observe in the world—can have a much lower degree of symmetry. Take, for example, a magnet. It picks out a particular direction in space and thereby breaks the perfect rotational symmetry of space, in which one direction is equivalent to any other. The basic law of magnetism is rotationally symmetric—all directions are equivalent—but a particular solution, a given magnet, breaks this symmetry by picking out a given direction.

This idea of a broken symmetry—a real system whose symmetry is lower than the particular law that governs it—was first used with great success in studying the physics of metals and other solids. It soon found an application in elementary particle physics when physicists realized that real particles break the symmetry of an underlying law in their lowest energy states.

This phenomenon is also referred to as the broken symmetry of the ground state or the vacuum. The basic idea is that some general symmetry is broken or hidden in the lowest-lying energy states, also called the vacuum state, but the full symmetry can still be revealed at high enough temperatures or energies. Here, in the case of elementary particles, the term *high energy* implies energies far beyond those we normally deal with, comparable to those energies involved during the first instants of the creation of the universe.

In the case of the magnet, the rotational symmetry of its basic laws is broken when the magnet is in its normal low-energy state, for its field picks out a particular direction in space. But heat up the magnet, excite its quantum components, and it will soon lose its magnetism. In so doing, it no longer picks out any special point in space. The broken symmetry is now restored; space is totally symmetric.

Physicists speculated that exactly the same thing happens with the weak-electromagnetic force. According to the gauge field approach, the weak-electromagnetic force is already unified and carried by a massless gauge field of infinite range associated with four massless vector bosons. However, in this force's lowest energy states, its basic symmetry or unity is broken. The weak and the electromagnetic forces now assume different strengths, three of the four gauge particles acquire masses, while the third, the photon, remains massless. It turns out that the two gauge field vector bosons W^+ and W^-, as well as the neutral vector boson Z°, assume quite large masses. The result is that, rather than having an infinite range, the weak interaction is now of extremely short range. However, at the very high energies found during the first instants of the creation of the universe, these two forces of nature would appear identical.

Physics had therefore been able to unify two of the fundamental forces of nature in terms of gauge fields whose underlying symmetry is broken at low energies. The great triumph of this theory was the eventual discovery of the gauge field vector bosons themselves. The first experimental indications of a W particle were found in 1982, and the Z was discovered on May 4, 1983. The detection of these vector bosons was to win Carlo Rubbia the Nobel Prize.

Physicists were also able to apply the ideas of gauge fields to the glue force acting between quarks. One of the great headaches of the 1950s, the strong nuclear force, was now seen in an entirely new light. Instead of trying to account for this unmanageable force in terms of the exchange of mesons between protons and neutrons, physicists viewed it in terms of gluons, the elementary particles of the gauge field, being exchanged between quarks. Since protons, neutrons, and the other hadrons are composed of quarks, the overall effect of this manageable gauge force becomes, in the large scale, a reproduction of what was earlier taken to be a fundamental force of nature—the strong force.

Could the gluon force between quarks be unified with the already unified electroweak force? Many physicists believed this was possible. Essentially there would then be a single force within the atomic world, a gauge force carried by massless vector bosons. The new symmetry of this unified force would be made by combining SU(3)—corresponding to the gluon force—with SU(2) × U(1) for the electroweak force. The product of these two symmetries becomes SU(3) × SU(2) × U(1). During the first instants of the universe, such a force is unified and described by some more general symmetry—(at the time, physicists assumed this would be SU(5). However, as the universe began to cool, SU(5), the basic symmetry of this fundamental unified force, would break into two parts, SU(3) and SU(2) × U(1). Next, the gluon force, described by SU(3), begins to separate from the electroweak. A further symmetry breaking would then take place as three of the vector bosons of the electroweak force began to acquire mass and electromagnetism became distinct from the weak nuclear force. SU(2) × U(1) now breaks into two separate groups. This was the aim of grand unification: nothing less than a combination of the standard model of the elementary particles—six basic quarks and six leptons—with a universal force.

The next step in the search for the Holy Grail was to explore additional symmetry schemes, mathematical mirrors that could not only reflect the hadrons and leptons separately among themselves but also reflect leptons into hadrons and vice versa. Within such grand unified schemes, all the elementary particles could be generated out of a single great symmetry, and all the forces of nature created out of a single gauge field. With SU(5) as the overarching symmetry it was possible to account for certain mysteries, such as the fact that the size of the charge of the electron is equal to the charge of the proton, and all the quarks and leptons certainly did fit into its symmetry pattern. But the new scheme also proposed new bosons with enormous masses—10^{15} times bigger than the mass of the proton. While these new bosons would be far too massive ever to be observed in a particle accelerator, their existence would have a curious implication.

Up to now a basic law of particle physics, known as baryon number conservation, had been connected with the observation that the proton does not decay. The proton was the most stable thing in the universe; it could never decay, its lifetime was infinite. But now with the new grand unified theories, their interparticle democracy, and those new massive bosons, it was possible for the proton to decay. Admittedly the lifetime of the unstable proton was a considerable 10^{33} years. The proton was, on average, going to outlive the universe itself. However, take a large lump of matter containing some 10^{33} protons. You would expect to see one or two disintegrating protons within this mass each year.

Physicists began to set up experiments to look for disintegrating protons, locating their apparatus at the bottoms of the deepest mines in order to filter out unwanted cosmic ray particles. But despite a number of careful experiments, there has as yet been no good evidence for the disintegration of the proton. One of the most important predictions of grand unification has yet to be confirmed, and this lack of observation of decaying protons was beginning to make physicists very uneasy.

One solution was to reject SU(5) and attempt a new grand unified theory using the group SU(10). But one implication of this approach would be that the normal left-handed neutrino now has a right-handed partner. The mass of the left-handed neutrino would no longer be zero, and its right-handed cousin would have an enormous mass—another problem for particle physicists. Yet even at this juncture, scientists were planning yet another step toward unification, one they hoped would bring even the gravitational force into the picture—supersymmetry.

Supersymmetry

The symmetry schemes of particle physics attempt to relate the elementary particles to each other in terms of reflections in symmetry mirrors such as isospin and strangeness. With the grand unified theories, it seemed that all the particles of matter—hadrons and leptons—could now be related and reflected one into another. It turned out that these idealized reflections were not quite realized in nature, and this failure had to be explained in terms of symmetry breaking. Nevertheless at high enough energies, the theory claimed, these particular particles will look identical; it is just that their basic symmetry is broken at low energy.

There remained one outstanding distinction within the elementary particles, and that was between fermions and bosons. You will recall that while the fermion family accounts for all the elementary particles of matter, such as fermions and leptons, the bosons include the force particles such as the photons, gluons, and W and Z particles of the weak force. The fermions all have fractional charges, while bosons have integer charges.

One aim of supersymmetry, which surfaced in the late 1970s, was to relate bosons to fermions. Within the mirror of supersymmetry, bosons become fermions and vice versa. It was to be the final great unification of physics.

It turned out that this new symmetry was essentially different from the other abstract symmetries of particle physics. To begin with, the symmetries of everyday objects are expressed in our three-dimensional space, while isospin, strangeness, etc. require an abstract internal space. Supersymmetry, however, involves a marriage between both spaces, for its mirror is partly in our conventional space and partly in an abstract space. This in itself was a particularly interesting theoretical move, for it suggested that there may be a way of combining abstract symmetries with operations in space-time. In fact, the supersymmetry had certain conceptual origins in string theory. In trying to create a string theory that allowed for fermions, Joel Scherk had eventually come up with a formalism that combines fermions and bosons within the same string. Particle physicists in the late 1970s were quick to embrace this formalism, while forgetting about the string theory that went with it.

Supersymmetry was the final icing on the cake of unification. Now fermions and bosons could be transformed one into the other, and all particles could be gathered into a single great family. Admittedly there were problems with this new approach. To begin with, it created a range of new elementary particle partners. When the photon saw itself in the supersymmetry mirror, it became a half-spin photino. Likewise, the other force particles became fermions—winos and gluinos and gravitinos. Quarks and leptons for their part became squark and slepton bosons. But where were all these new elementary particles? Experimentalists had never seen them; did this mean that they were all so heavy as to be beyond the range of even the largest particle accelerators?

One way out of this quandary was to propose that supersymmetry is yet another of those broken symmetries of physics, so that the various supersymmetric partners are no longer of equal mass. In fact, the new leptinos, photinos, gluinos, and so on will have very high masses, so high that they need not show up in particle accelerator experiments.

The new theory opened a further possibility: What if supersymmetry is a local symmetry? If this were true, then physicists would have to work out the nature of the corresponding gauge field. The result was fascinating: a field with a massless gauge particle having spin 2. In this case, physicists did not need to propose a breaking of symmetry. They were very happy with a massless, spin 2 vector boson, for this was none other than the graviton, the elementary particle of the gravitational field. The local symmetry that reflects fermions and bosons into each other is nothing less than the gravitational field! If this is true, then it becomes possible for gravity to enter into particle physics along with the other forces of nature as a gauge field. And will this supergravity, as it came to be called, unify with the three strong and electroweak forces? Physicists immediately began to work on a new, unified supergravity theory.

The picture presented by grand unification and supergravity was breathtaking. The only problem was whether it was true. Does nature really behave in this way? Do squarks and photinos exist? Are all the various gauge particles real? Do the patterns of grand unification and the standard model of quarks have any correspondence to reality? Is the great program of unification a reality, or is it a fantasy, an illusory vision that had swept physicists away with its claims?

By the early 1980s many physicists were becoming disillusioned with the whole approach. Not only were so many new, and as yet unidentified, particles demanded, but the various symmetry schemes were arbitrary. Each school of unification had its own symmetry approach, its own pattern for the elementary particles, and there seemed to be no way of deciding between the various proponents. Physicists also pointed to the enormous difference in energy that existed between what could be detected experimentally and that vast, unexplored region before the broken symmetries are theoretically restored. For example, the symmetry between electromagnetism and the weak nuclear force is supposed to be restored at energies of 100 GeV. Now, to reach this with human-made particle accelerators is not out of the question; it may well be eventually possible to probe this unified force experimentally. But what about unification of the electroweak force with the gluon force? This requires an unimaginably high energy (see diagram). It is hard to imagine physicists ever being able to probe such a region—an energy that happened within the first 10^{-38} seconds of the birth of the universe. What meaning is there to a theory that permits such an enormous energy gap? After all the energy range between electroweak and grand unification is equal to that between our own scale of things and that of electroweak unification—and look at the rich physics that has been uncovered in this energy range during this century. Who knows what unimagined phenomena will be detected beyond the electroweak energies before we reach the hypothetical energy at which grand unification is finally achieved?

By the early 1980s physicists were discontented. Matters had not gone as well as they had hoped. The initial excitement of the grand unification approach had worn off. What was needed was some dramatic new idea, a move in a totally new direction. In 1984 that direction was signposted. Its name was superstrings.

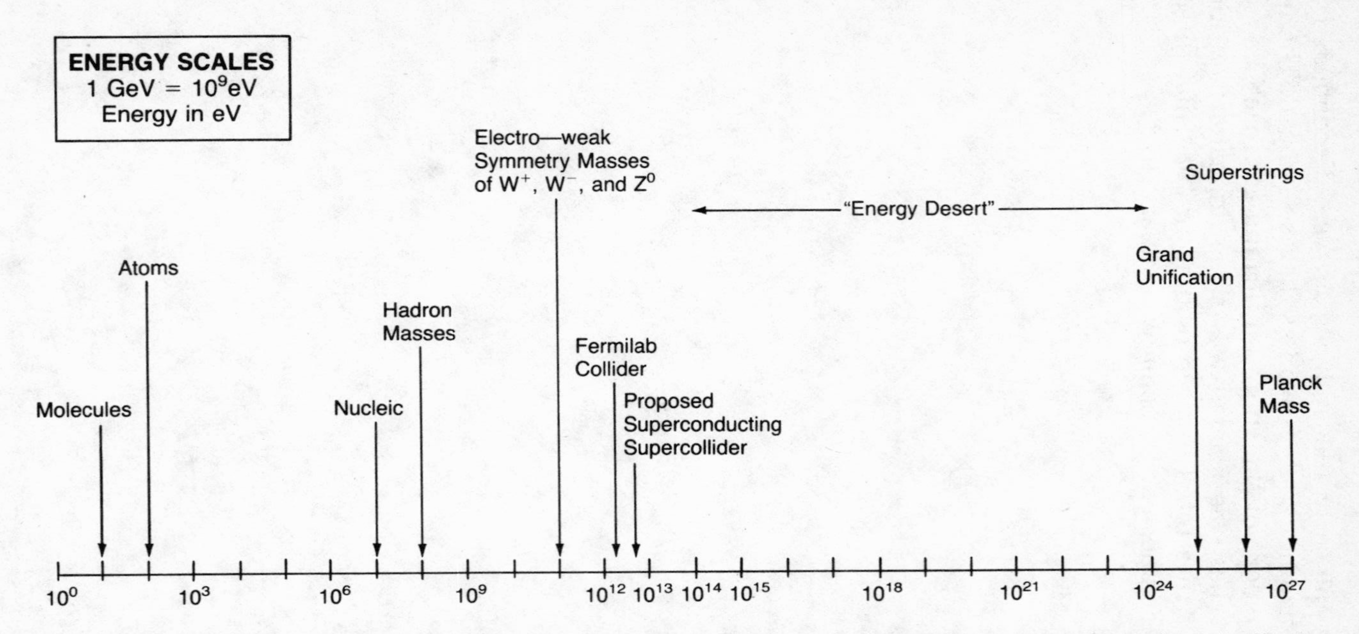

Note the vast range of energies beyond the limits of present particle colliders and the new physics of superstrings and grand unification.

5 Superstrings

THE GRAND UNIFIED theory was supposed to be the great success of contemporary theoretical physics, the Holy Grail that would finally be grasped after decades of searching. Not only would the hadrons and leptons be gathered together into one great symmetry pattern, but various forces of nature would be similarly unified. Already electromagnetism and the weak nuclear force were known to be the result of a single, electroweak force. In turn, at energies that existed during the creation of the universe (that is, at distances around 10^{-33} cm or within the first 10^{-38} sec. of the birth of the universe), this electroweak force would unite with the strong nuclear force.

When supersymmetry is added to this grand unified theory, then all particles, fermions and bosons, become unified within a new space called *superspace*. Finally, if supersymmetry is viewed as a local rather than a global symmetry, it is necessary to add a gauge field. It turns out that the gauge field of supersymmetry is the gravitational field. Gravity is added to the grand unified force and all of physics becomes one at 10^{19} GeV.

This was the grand design that was being put into place at the end of the 1970s. The only problem was that it left much to be desired. Where were all those extra particles demanded by the theory, in particular the supersymmetric partners to the known fermions and bosons? And how was the theory to be made internally consistent? It seemed to be plagued with many of the formal problems that had haunted particle physics for decades. Finally, if this was truly the grand scheme for unification, then shouldn't it be unique? Shouldn't there be just one way to unify physics? But physicists were proposing a wide variety of alternative symmetry schemes, and there seemed to be no good way of choosing among them.

If grand unification was the Holy Grail, then, like that grail sought by King Arthur's knights, it had the habit of vanishing just when it appeared within grasp. Even more distressing, there were many grails, each with its own claim to authenticity.

By the early 1980s, physicists were feeling uneasy about the whole approach. What seemed to be needed was a vital new idea, a novel approach that would lead, uniquely, to a single consistent theory about subatomic matter. It was into such an atmosphere of dissatisfaction that Michael Green and John Schwarz burst their bombshell—a theory of the elementary particles that was free from infinities and anomalies, a theory that was consistent and unique. Its name was superstrings.

By reformulating the original string theory of the early 1970s and adding to it the new insights of supersymmetry and grand unification, Green and Schwarz had come up with an approach that not only described hadrons, as the original string theory had attempted to do, but *all* the elementary particles. Even the forces of nature emerged from the theory in a totally natural way, as a consequence of the geometry and topology of strings. And since calculations using the theory did not blow up but produced finite predictions, and its symmetry pictures were unique with no room for arbitrary factors, some physicists not surprisingly hailed it as *the theory of everything*.

But how exactly did Schwarz and Green come to create their superstring theory? John Schwarz had been one of the early workers in the field of Nambu's original string theory, but by the end of the 1970s, he had also developed an interest in supergravity and was now working in that field. Michael Green, however, was still interested in the old string theories. Then one day the two physicists happened to meet by chance while on a visit to CERN, the European Center for Nuclear Research in Switzerland. Over coffee they began to talk about their respective interests and the possibility of a connection between string theory and supergravity. A year later, in 1980, they had formulated their first version of the new string theory.

To begin with, they realized that a string is not really a theory of the hadrons, which implied strings some 10^{-13} cm long. In fact, it is a theory about *all* the elementary particles and *all* the forces of nature. A major clue as to why this must be so had been around in the days of the old string theory, when physicists had noted that closed strings looked like massless bosons with a spin of 2. But such massless particles would have to be the quantum particles of a gravitational field. Perhaps, Schwarz had speculated at the time, strings not only describe hadrons—the strongly interacting elementary particles—but force particles such as the graviton as well. As he was to write later, "String theory requires the existence of gravity, whereas the point-particle theories require that it does not exist. This consideration sustained my enthusiasm for the subject throughout the ten-year period prior to its widespread acceptance."

But since strings describe everything—particles and forces—theoreticians reasoned that they could not be as long as 10^{-13} cm; they had to be far, far shorter, only 10^{-33} cm. If strings are to represent gravitons, then their scale must be around that at which space-time begins to break down. So Schwarz and Green began with very short strings indeed, strings that, moreover, were specifically designed to exhibit the full supersymmetry that links fermions and bosons. The new *superstrings* would, they hoped, apply to all of nature's quantum particles alike.

Michael Green and John Schwarz decided to use their summer vacations away from teaching in order to meet and work together. They progressed quite rapidly, and a year later, in 1980, they had developed a supersymmetric theory for their new, tiny strings. They called this first version the Type I theory, and it described open strings moving in ten dimensions. While pictorially these strings may have looked like minute versions of Nambu's first strings, joining ends and breaking apart again, nevertheless they were profoundly different. The new Type I strings were supersymmetric and combined the symmetries corresponding to both fermions and bosons. In addition, the aim was to describe all the elementary particles and forces as emerging out of this much deeper theory.

A year later Green and Schwarz had put together a different version of the theory, called the Type II theory, in which the strings now came in the form of closed loops, again moving in ten dimensions, with fermions and bosons running around the loop in opposite directions. But before we unfold the various phases of Green and Schwarz's research, let us see in a little more detail exactly how a superstring is created.

Creating a Superstring

The process of generating superstrings requires a series of careful steps:

1. Discovering the correct mathematical equations that describe the way a string moves and vibrates
2. Ensuring that these equations conform with the theory of relativity
3. Quantizing these equations of the relativistic string
4. Ensuring that the strings are supersymmetric and that they will also relate to one of the various symmetry groups of the known elementary particles

In fact, there should be an additional step, which physicists call second quantization. Step 4 ends up giving us the wave function of the string. Second quantization takes this even further and involves creating what is known as a quantum field out of these wave function solutions. In a sense second quantization, or *quantum field theory,* is a much deeper way of doing things. On the other hand, a second quantization poses considerable technical difficulties, which have yet to be resolved. In Green and Schwarz's case, the two physicists were interested in going only as far as the fourth step to produce a consistent theory of quantum superstrings that is free from all infinities and anomalies. The possibilities of an additional step to produce a full superstring field theory will be discussed in the final chapter.

Let us look at these steps in turn. Schwarz and Green had to get them exactly right in order to create their superstring, and even then their problems would not be over.

Step 1: A Vibrating String

Their first task was to begin with a string of the right length and tension. Since the theory was supposed to account for, among other things, the quantized force of gravity, the strings could not be longer than 10^{-33} cm. The tension in the string was chosen as an immense 10^{39} tons to accord with other reasonable estimates.

The string itself is specified by a series of coordinates or labels, which refer not only to its various configurations but also to its internal symmetries. These symmetries have to be large enough to include all the insights gained by the grand unified approach. In fact, the way in which this symmetry is eventually chosen becomes a high point in the story of superstrings. The major problem at this stage, however, is how to describe the various movements, twists, rotations, and vibrations of this one-dimensional object. How is this to be done?

The string can twist and turn, spin and vibrate, tie and untie itself, do anything the human mind can imagine. But there must be some restriction upon this string, some mathematical way of ensuring that these various motions conform to the laws of nature. You may recall from school that Newton's laws are supposed to describe the motion of material bodies. Given the mass of a body and the forces acting on it, we can discover just where and how it moves. In the case of ordinary particles, Newton's laws are conveniently used to discover their overall motion.

But there is another formulation of Newton's basic laws of physics, another way of arriving at the same solution. In fact, this alternate approach is very useful when physicists want to make the translation from classical mechanics into quantum theory. In the present case, it is done by considering the motion of the entire string at once and is called the principle of least action.

It turns out that this principle is far older than the laws of Newton. In fact, Hero of Alexandria in 125 B.C. proposed that light always travels between objects by the shortest path—an example of a principle that applies to the whole of a path, rather than the piecewise laws of Newton. Leonardo da Vinci had realized that the same principle also applies to a falling body, and Gottfried Leibniz had suggested that nature always operates upon the principle of maximum economy. (This idea, that nature selects the "best of all possible worlds," was satirized by Voltaire in his book *Candide*.) Finally, in the eighteenth and nineteenth centuries, physicists like Leonhard Euler, Joseph-Louis Lagrange, and William Rowan Hamilton put the principle on a firm mathematical footing. They argued that, out of all possible paths or motions, nature selects the path that minimizes what they called the "action." (*Action* is a technical term and could be thought of as derived from the momentum of a particle.)* This action is not, however, evaluated at a

---

**Action* can be defined as $\int_{t_1}^{t_2} pv\,dt$, that is, an integral or summing up of the momentum p and velocity v of a particle between some initial and final time.

single point, but is summed up over the whole path of an object's motion.

In other words the principle of least action dictates that, out of all conceivable motions, or paths, physicists must pick the one that minimizes the total action. To put it in terms of a mathematical prescription: Write down a general expression for the action of a system, then choose the path or form of behavior that will minimize this total action. The result will give the actual behavior of the system. While in many cases Newton's laws are a useful way of proceeding, in others it is easier for physicists to use this least-action approach.

In the case of a classical string, physicists would begin by writing down the general expression for the string's action, and then, using a straightforward mathematical generalization of the calculus, they would minimize this mathematical expression. The result describes the infinity of legitimate vibrations and rotations of a classical string.

Step 2: The Relativistic String

In his special theory of relativity, Einstein had demonstrated a number of bizarre effects: that masses increase, clocks run slower, and measuring rods shrink when physical systems move at speeds close to that of light. An immediate consequence of this is that the same phenomenon will have different appearances to observers who move at different speeds. However, Einstein had stressed the importance of an underlying unity to nature; no matter how varied these appearances may be, the same underlying laws must hold for all observers. Einstein was saying that the mathematical meaning of a physical law must not change as we go from one observer to another.

Now each observer has his or her own perspective on space, in more technical terms: the person's own coordinate system. In our everyday world, we tend to use the same maps, but Einstein taught that when we move at speeds near to that of light we must each carry our own individual maps (or coordinate systems) of the world around us. In going from one point of view or observer to another, we must therefore exchange maps or coordinate systems. You have your map or coordinate system, and I, moving at a very different speed, have mine. When we compare our observations, we translate between maps or coordinate systems. But while these map changes—or coordinate transformations, as mathematicians call them—change the appearance of a phenomenon, *they must leave the basic form of nature's laws unchanged.* This means that the form of the underlying law should not change, even when written in a different coordinate system. While the appearances of things may depend upon an observer's state of motion, Einstein taught, the underlying laws of nature must be identical for all observers.

In the case of a relativistic string, this means that the action has to be written down in such a way that in going from one coordinate system to another, from one observer to another, its physical form remains the same. In physicists' terms, the action must be *manifestly covariant.*

It turns out that this requirement of manifest covariance severely restricts the equations of a superstring—particularly when those equations eventually come to be quantized. But this was exactly what Green and Schwarz wanted, for the more restricted a theory becomes and the more it is hedged in by mathematical demands, the less arbitrary can be the choices made by physicists themselves. A perfect theory would be forced on physicists by nature itself; there would be no room for arbitrary assumptions or for making adjustments. The theory would stand or fall on its own.

Going from a *classical* to a *relativistic* string involves a significant change of description. To begin with, the notion of the path of a string is very different in a relativistic theory. Conventional nonrelativistic theories involve writing down equations in space and then discovering how the system develops or evolves in time. Time and space are treated as separate. A relativistic theory takes a very different approach. In 1908 Hermann Minkowski, Einstein's former mathematics professor at the ETH (Swiss Federal Polytechnic School) in Zurich, gave an address to the eightieth Assembly of German Natural Scientists and Physicians at Cologne. It began, "Henceforth space by itself, and time by itself, are doomed to fade away into mere shadows, and only a kind of union of the two will preserve an independent reality." The notion of space-time was born, and with it a new conception of the trajectory of a particle.

Imagine yourself floating in outer space and trying to describe the position of various stars as up or down. Another astronaut floating at an angle beside you will have a different sense of up and down. Both of you are free to make choices about your convention of up and down—you have a local symmetry with respect to the directions of space. But as soon as you begin to talk to each other by radio, then it becomes necessary to translate from one convention to the other. In mathematicians' terms, each observer has his or her coordinate system, but in order for the observers to agree on what they are seeing, they must translate between coordinate systems—in this case make rotations, between each person's axes of reference. In other words, your up and your right become mixed together when seen from my coordinate system.

Figure 5–1
These creatures, floating in space, do not agree on which way is up and which is down.

In his ground-breaking paper on special relativity, Einstein showed that when observers are moving at different speeds, they need a new set of coordinate translations that involve not only spatial directions but the temporal direction as well. (These effects are only important at speeds near that of light.) In other words, when one translates between moving observers, spatial coordinates will become mixed up with the time coordinate.

As a mathematician, Minkowski realized that this result would be more elegant if the time coordinate was brought in, from the start, on an equal footing with the three spatial coordinates. In this way, the four dimensions of space-time were born. An immediate consequence of this new formulation was the way in which physics now talked about the movement of material bodies. In the prerelativistic picture, a stone can be considered as a point in space. A stationary stone keeps the same position in successive times, while a falling stone changes its position from one instant to another. In relativity, however, the stone's description is given not in space alone but in *space-time*. In place of the point there is a line, called a world line, a point that has been dragged out into the fourth dimension of time. This world line could be thought of as containing the entire history of the stone at all its different points in space and for all times. But this has a very important consequence for the relativistic action principle, for it must minimize the length of this world line.

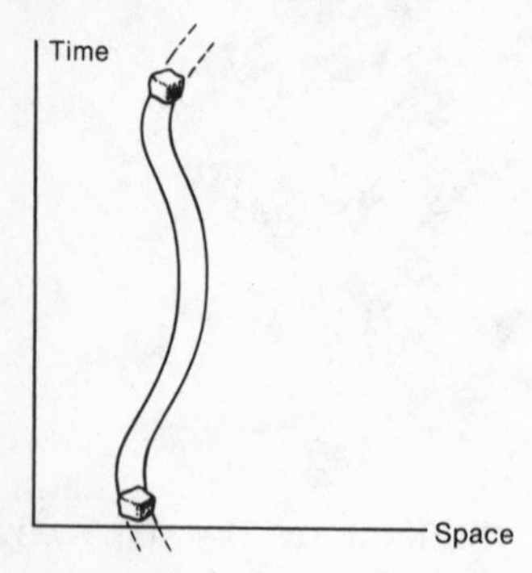

Figure 5–2
The world line of a stone indicates every position occupied by the stone in space during its lifetime.

Now back to the relativistic string. In our conventional space, a string is a one-dimensional object. But in space-time it acquires the additional dimension of time. As a result, the string becomes a one-dimensional line that becomes swept out in the fourth dimension of time. In other words, a string becomes a two-dimensional surface. Not a world line but a *world surface*.

Figure 5–3
The world line of a string forms a two-dimensional surface in space-time.

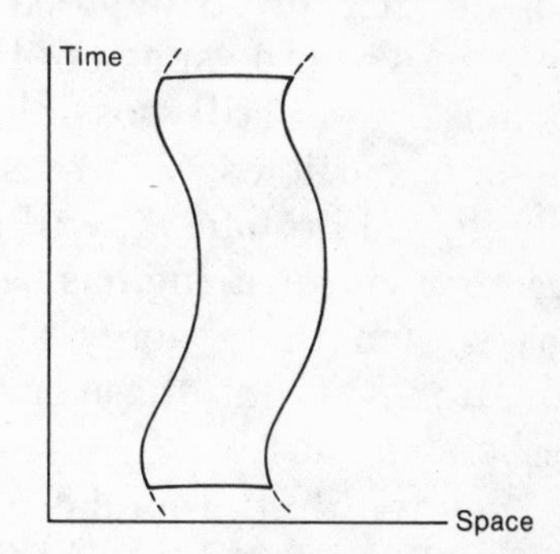

The relativistic action principle now demands that the area of this world surface be minimized. In other words, strings in our three dimensions of space could be thought of as temporal snapshots, instantaneous slices in time of these minimal surfaces in four-dimensional space-time.

It turns out that the concept of a minimal surface is already a familiar one. You probably blew minimal surfaces as a child and will certainly have observed them while washing the dishes or having a bath. Minimal surfaces are those shapes made by soap bubbles and soap films, for these films attempt to minimize their energy by having the smallest surface areas. The same branch of mathematics that is used to study soap films also determines the minimal surfaces that correspond to relativistic superstrings—those strings which correctly satisfy the demands of a relativistic action principle.

To sum up, relativity imposes severe restraints on the theory of strings. It insists that in going from the perspective of one observer to the next, the law of the string should not change. To put it mathematically, there is a relativistic symmetry within the theory which dictates that the basic action principle must be covariant (essentially unchanged) by transformations among its coordinates. The implication of this relativistic action principle is that strings correspond to minimal surfaces in spacetime.

The immediate consequence of these relativistic restrictions is that the string must be massless and that its ends must move at the speed of light. Yet, despite this absence of mass, it is still possible for this string to represent massive particles. The reason is that because the string vibrates and rotates, it will have a series of energy levels, just as a violin string has a series of notes. Since Einstein has shown that energy and matter are related by the equation $E = mc^2$, these energy levels therefore also have associated masses.

Even within these relativistic restrictions, there are additional symmetries, additional transformations among the coordinates which must leave the basic physics of the relativistic string unchanged. Suppose, in the following diagram, we interchange the coordinate P on the world surface of the string with the coordinate P' outside. This has the effect of deforming the minimal world surface and therefore of changing the value of the action. But since the action is already a minimum, no deformation is allowed, for that would only increase the value of the action. A proper string theory does not allow us to relabel coordinates in this way.

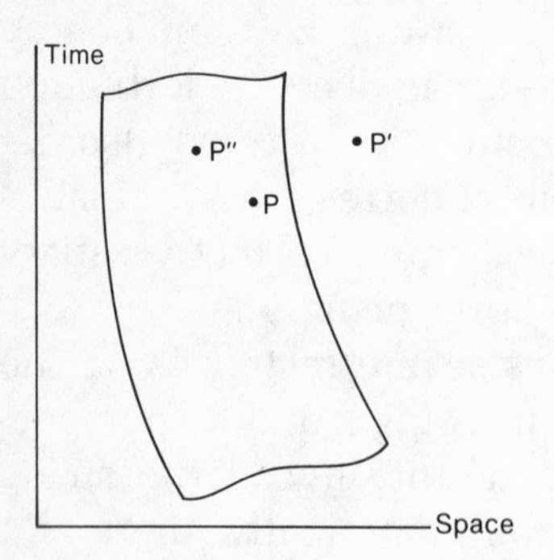

Figure 5–4
Interchanging coordinates P and P'' which lie on the world surface of a superstring does not change the value of its *action*.

But suppose we interchange *P* with *P''*. Both coordinates are within the surface itself. From outside, nothing would appear to have changed; the action would be exactly the same. A proper superstring theory must therefore allow this particular symmetry among the coordinates or labels of the string. This may sound trivial; however, it does restrict, in a very interesting fashion, the possible ways in which the relativistic action can be written down. When we come to quantize the string, these coordinate symmetries restrict the possible mathematical moves, pointing physicists in the right direction.

Steps 3 and 4: The Quantized, Supersymmetric String

These final steps in this stage of the superstring story are to quantize the relativistic string, obtain a complete description of its various quantum states, and ensure it has the correct symmetry. But it was at this exact point where all the problems began to manifest themselves in the earlier versions of string theory—tachyons, infinities, ghost states, and the like.

Just as in going from a classical to a relativistic string it is necessary to write down the mathematical formalism in exactly the right way, so too quantization requires an expression that has the correct form and symmetry. In fact, there is a definite prescription for quantizing a relativistic string. The equations governing the string are first written down in this standard form, and then certain terms are replaced by their quantum mechanical equivalents. Specifically, terms labeling the string behavior that are expressed in terms of coordinates are replaced by what are known as quantum mechanical *operators*.

A quantum operator corresponds directly to a measurement of the quantum system itself. For example, take the coordinate x, which is used in the nonquantum world to label the position of a particle. In our everyday world, if we wanted to know the location of a particle, we would ask for its coordinate, x. However, in the quantum world, this location may be indeterminate or probabilistic. The best we can do is to measure its position and read out the result; a further measurement will give a slightly different result, and so on. So, rather than using precise labels, the quantum theory deals with operators that are in a sense the mathematical representations of making such measurements or asking a quantum question. The result of a particular quantum measurement or experiment then becomes equivalent to applying a quantum operator. The answer may be a definite state, or it could be a probabilistic result.

In classical mechanics we begin by writing down a law of motion—Newton's law or an action principle. This law takes the form of a differential or an integral equation, and its solution gives a "readout" of the coordinates, that is, the positions of the particles at different times. But in quantum theory, things don't work this way. Rather, quantum operators are mathematically applied to the system in order to get solutions. This is, as was explained in the previous paragraph, equivalent to carrying out a quantum measurement. The solutions—the results of applying the quantum operator—now correspond to different quantum states (wave functions), each with its own characteristic energies, quantum numbers, and symmetries. To carry out this program, therefore, Green and Schwarz first had to take a correct relativistic description of the string and then use the prescription of replacing certain terms with quantum operators.

In the case of the early string theories, once this procedure had been carried out, the solutions, or wave functions, were found to involve physical absurdities. There seemed to be no way in which Nambu's theory was able to select between what was physically reasonable and the absurd unphysical solutions.

But now, in their superstring approach, Green and Schwarz were adding ever more restrictions to their theory. Experience suggested that all the problems about unphysical solutions and anomalies should get far worse when the quantization step was taken. Superstrings didn't look as if they stood a chance. But the two physicists felt differently. Making all the correct choices and exploiting the right symmetries, they argued, will leave less room to maneuver. Either quantized superstrings will be right on target, or the results will be absurd and the whole theory will have to be thrown out.

In taking their final step of quantization, Green and Schwarz had to impose yet another symmetry upon the supersymmetric theory. It was called chirality, and it expresses the basic handedness of nature.

Chirality: Left-Handed Versus Right-Handed

The history of physics over the last two decades could be thought of as first uncovering basic symmetries and then finding out that, in reality, nature breaks these symmetries. (If this sounds as if scientists have got things the wrong way around, then Roger Penrose would agree with you. His rejection of this philosophy of symmetry breaking is discussed in Chapter 9.)

While the breaking of nature's symmetries has occurred again and again, there were symmetries that appeared so fundamental that physicists believed they could never be broken. One of these was time reversal symmetry, the idea that the laws of nature do not involve themselves with the direction of time, or "time's arrow" as it is called. Another intrinsic symmetry was that between right- and left-handedness. When nature's particles are reflected in a mirror, there should be no reason to choose the right- over the left-handed.

Of course, in our own world, right- and left-handed people are different. In fact, handedness even comes in at the level of important biological molecules—left-handed molecules enter into biological processes, while the chemically identical right-handed molecules do not. But physicists believed that, at the quantum level, this handedness must be a perfect, unbroken symmetry. In a nuclear disintegration, for example, 50 percent of the particles will come out with right-hand spins and 50 percent with left-hand spins. Nature will not prefer one hand over the other. Imagine, then, the shock when, in 1957, under the influence of the weak interaction, the decay of the K-meson was discovered to behave in an anomalous way. At first sight, it looked as if there were *two* K-mesons, identical in every respect yet differing in the way they could decay. It was Tsung-Dao Lee and Chen Ning Yang who suggested that there is only one K-meson but that, when it comes to weak interaction decay, the left-handed spinning meson behaves in a different way from its right-handed partner.

The very idea that nature would prefer one form of handedness over another seemed absurd. Wolfgang Pauli, a physicist of great physical intuition, was willing to bet a sum of money that God is not left- or right-handed but always preserves this essential symmetry of nature. But only months after Lee and Yang's hypothesis, a group led by C. S. Wu discovered that electrons emerging from a decaying nucleus of cobalt (again involving the weak interaction) were spinning in a preferred direction. God was left-handed and Pauli had lost his bet. "The worst mistake of my life," Pauli is reputed to have said at the time, and then to have recovered by claiming that his mistake lay not so much in his intuition but in his willingness to bet money on it.

From now on, physicists would have to ensure that any theory which included the weak interaction, or the electroweak interaction as it should now be called, must exhibit the correct choice of handedness. This property was termed *chirality*.

Making sure that a theory ends up with the proper chirality had been a major problem for string theorists. It was enormously difficult to satisfy the restraints of supersymmetry, quantum theory, and relativity and at the same time guarantee that chirality is retained. Even when things looked as if they were working out in higher-dimensional spaces, chirality seemed to fly out of the window as soon as physicists condensed their results down to our three spatial and one time dimension.

Although string theories are formulated in higher-dimensional spaces, their results eventually have to be translated to our more familiar space-time of one time and three spatial dimensions. This is done by a process called *compactification,* in which the extra dimensions are "rolled up" like a carpet. But this act of compactification generally destroys the chirality or handedness that had been built into the theory. Chirality and the other requirements of a good superstring theory seemed to be incompatible. The solution to this problem was to become a major goal for Green and Schwarz.

How Many Dimensions Does the Theory Need?

Physicists working on point particle theories had been forced to look to work in higher-dimensional spaces when they discovered that supersymmetry was impossible to apply in our four-dimensional space-time. Superstrings present similar difficulties. In the case of point particle theories, Edward Witten, another of the superstars of superstrings, had presented a strong argument that eleven dimensions were the natural space in which to represent the gauge forces of nature. The strong and electroweak interactions can only be properly recovered from point particle theories in a space that has either six, ten, or eleven dimensions. On the other hand, if the space is made smaller than eleven dimensions, all sorts of other problems apparently would crop up. It appeared that eleven dimensions were forced on physicists as the space of choice for the grand unified supersymmetry theories. Eleven dimensions therefore seemed a natural choice for all point particle theories. But did this necessarily apply to superstrings?

Witten had also proved that a chiral theory is only possible in a space having an even number of dimensions. Nature's handedness could not show up in the odd number of dimensions of an eleven-dimensional space. But even if physicists decided to work in a ten-dimensional—an even-numbered—space, they could still not guarantee the basic handedness of nature. It was possible to create a chiral theory in ten dimensions, but as soon as six of the ten dimensions were rolled up or compactified, to leave the remaining four dimensions of space-time, chirality was lost. Thus, not only was this chirality difficult to produce, it was almost always destroyed by compactification.

These are just some of the problems that Green and Schwarz had to face when, in 1979, they decided to press ahead with a new theory of superstrings.

Closed or Open Loops

The natural space in which to formulate a string theory is ten dimensions, which allows for chirality or handedness. But would this chirality still be around when the superstring space was rolled up or compactified to produce four dimensions? And how were the various gauge forces of nature to be recovered from this ten-dimensional string theory? Without a proper account of these gauge forces (also called Yang-Mills forces), the theory would make little contact with the physics of the real world.

Nevertheless, in 1980 Green and Schwarz were able to formulate a supersymmetric theory of open strings, or superstrings. This was called a Type I theory, and a year later they produced a second version, called a Type II theory, which this time involved only closed loops.

In the Type II theory, fermions and bosons were envisioned in an ingenious way as waves that move around the closed loop of the superstring. Fermions travel around the loop in one direction, while bosons move the opposite way. The closed loop is therefore truly supersymmetric, for it contains bosons and fermions on an equal footing.

Closed-loop superstrings can have quantum numbers that are identical with those for the quantum particles of a gravitational field—the gravitons. In other words, by 1981 Green and Schwarz had created a theory that seemed to give a good account of what point particle theorists were calling supergravity. It also triumphed by being a finite theory, for when Green and Schwarz used perturbation theory to make calculations, the results did not blow up to infinity as soon as the physicists started on the mathematics. A finite theory was in fact one of those characteristics of the Holy Grail of physics!

But Green and Schwarz could not rest on their laurels, for it did not seem possible to recover the gauge forces of nature from such a theory. And what use was a superstring theory that could not explain the way in which the elementary particles interact?

The philosophy of the time was greatly influenced by what were called Kaluza-Klein theories, that is, theories in which the forces of nature appear when a higher-dimensional space is compactified or curled up. Theodor Kaluza and Oskar Klein had developed a five-dimensional theory of general relativity in which the act of curling up the extra fifth dimension produced what looked like an electromagnetic field. This electromagnetic field had not been put into the theory to begin with; rather, it emerged out of the process of compactification itself. Physicists who worked on point particle theories believed that it should be possible to make the gauge fields of nature also appear through a similar mechanism. Green and Schwarz's closed loop theory, although it was chiral and not plagued by infinities, did not seem capable of producing the various forces upon compactification.

Then, in 1982, Edward Witten and Luis Alvarez-Gaumé tossed a bombshell into theoretical physics. Their paper was essentially concerned with point particle theories and not with strings, yet its various conclusions influenced both approaches. One aspect of Witten's and Alvarez-Gaumé's approach had been to show that chiral theories are possible only in an even number of dimensions, such as ten. And even with chirality in a higher dimension, this basic handedness of nature can still be destroyed upon compactification.

But the 1982 paper also turned its attention to the philosophy of Kaluza and Klein, that the forces of nature are created through the compactification of some higher-dimensional theory. This sort of process, they showed, can happen only in an odd number of dimensions. If the forces of nature are to be created by a mechanism involving the curling up of dimensions, then physics has to begin in eleven, and not ten, dimensions. But, on the other hand, odd-dimensional theories could not be chiral. As the poet e.e. cummings once said, "you pays your money and you doesn't have a choice!"

Finally Witten and Alvarez-Gaumé alerted people to the possibilities of yet a new set of problems, called anomalies, which could not be tolerated in any well-behaved physical theory. These anomalies would involve, for example, the violation of powerful conservation laws, so that positive or electrical changes would appear to be created out of nothing.

The kiss of death for many theories came about when Witten and Alvarez-Gaumé showed that, while four-dimensional theories can be free of such anomalies, they will occur in two, six, and ten dimensions.

In creating a new theory of superstrings, Green and Schwarz had to work in ten dimensions, but did this mean that they had lost any chance of representing the forces of nature? And how could their theory avoid anomalies? Only a miracle could save them.

In the early 1980s, both Witten and Green and Schwarz knew about each other's work—preprints of important scientific papers are well circulated among scientists long before they appear in print. But now Witten and Alvarez-Gaumé had given Green and Schwarz a good reason to look for anomalies within their string theories. They had also provided a strong argument for why the gauge forces of nature need not be tied to the process of compactification—for such a process cannot work in an even number of dimensions. Green and Schwarz would have to have both the gauge and the gravitational forces already present in an anomaly-free, ten-dimensional string theory.

Let us look back for a moment at what Green and Schwarz had done. By the early 1980s, they had two alternative approaches to a superstring theory, a Type I theory involving open strings and a Type II theory using closed loops only. These theories possessed the essential handedness or chirality; moreover, they had been able to show that the closed-loop version did not blow up and give infinities.

However, it did not look possible to recover the gauge forces of nature from their closed-loop theory. From their perspective, in the early 1980s, a Type II theory seemed to lack the necessary power to generate the physics of the subatomic world. It could hardly be called a "theory of everything." [It has since turned out, however, that from the perspective of 1988, there may be new ways of looking at Type II theories and pulling out the gauge forces of nature. Many of the intuitions and ideas that were floating around in 1981 were strongly influenced by point particle theories and there seemed to be no way of obtaining gauge forces from a Type II theory. But now that physicists are thinking in more "stringy" ways, they are also suggesting that Green and Schwarz's Type II theories may not have been such a bad idea after all.]

Nevertheless, by 1982, Green and Schwarz decided to abandon their closed-loop approach and take another look at their first, Type I, open-string theory. Not only did such a theory have to be finite—that is, free from infinities—but, sensitized by Witten's and Alvarez-Gaumé's warning, they were on the lookout for anomalies, a major failing that seemed to be inevitable in any ten-dimensional theory.

Green and Schwarz had been trying to keep a phenomenal number of balls in the air as they juggled with superstrings: The theory had to be manifestly covariant in order to satisfy the requirements of relativity. It had to be quantized. It had to be chiral as well as reproducing the forces of nature and exhibiting the supersymmetry that relates bosons to fermions. It had to be free from anomalies. It had to account for the forces of nature.

In addition, the final theory, when viewed from a distance or at low energies, would have to reproduce the general appearance of the elementary particles along with the correct symmetry group necessary to produce their unification. There is no point in creating a superstring theory if it bears no relationship to what is known experimentally about subatomic physics. This final point meant that a particular grand unified symmetry would eventually have to emerge out of the superstring theory. But picking the correct symmetry had been one of those headaches that had been haunting conventional particle physics for the past decade. There turned out to be a host of potential symmetries. Their number is limited only by the ingenuity and imagination of physicists. Each symmetry produces its own pattern, or variant on a pattern, and physics seemed to have no real way of deciding between the merits of one scheme and another.

At first sight, the same problem was present in superstring theory. Once all those other balls had been successfully juggled, conventional wisdom dictated, there was still an arbitrary choice of symmetry and the goal of avoiding anomalies. But this meant that there was to be not one but a vast choice of possible superstring theories. Here John Schwarz and Michael Green parted with the conventional approach. Suppose, they said, this unifying symmetry is forced on us by nature. Suppose there is no room to maneuver and this unifying symmetry has already been selected for us.

This time, when Green and Schwarz met in 1984, they began to focus on the sorts of anomalies that would plague an open-string theory. In general it looked as if anomalies could never be weeded out of Type I theories.

But then a theoretical miracle happened.

Open-string theories were doomed to fail, except for a single choice of symmetry. When the theory was created to conform to what is known as an SO(32) symmetry, then all anomalies vanish. The theory is finite and free from anomalies; it also is rich enough to contain gravity and gauge fields of nature. It turns out that, for this one choice of symmetry, the gauge fields and gravity act together in such a way as to eliminate anomalies from the theory.

It was almost as if Schwarz and Green did not even have to bother with juggling this final ball. Provided that all the other balls had been kept in the air, nature itself took care of this final step and created a perfect theory. A unified symmetry was forced on the theory, and once this had been admitted, it became possible to create a superstring theory that had all the features physicists had been dreaming of. It was chiral, supersymmetric, and free from ghosts, tachyons, infinities, and anomalies. It accounted for the forces of nature and the symmetry patterns of the elementary particles.

The theory of everything had been born.

Within a matter of weeks, all of particle physics was talking about the new theory. To begin with, many theoreticians had been puzzling about supergravity and gauge fields. But now Green and Schwarz had placed these topics in an entirely new light and produced a theory of incredible power and attraction. In fact, Green and Schwarz had also gone on to show that some of these features would also be present in a point particle theory that uses either the symmetries SO(32) or E8 × E8. However, point particle theories no longer looked as attractive as string theories. From now on, an army of theoretical physicists would shift their focus from particle to string theories. A revolution in theoretical physics had occurred.

Forces and Strings

Before Green and Schwarz's revelations about open superstrings, the conventional wisdom was that the best place to look for new theories was in eleven dimensions. A supersymmetric point particle theory, formulated in a ten-dimensional space, for example, did not seem to have enough room to include the forces of nature. Gauge forces, it was assumed at the time, must arise out of the process of compactification as extra dimensions are curled up—and this simply would not work in ten dimensions.

But Green and Schwarz had suggested a radically new approach, for gravity and the gauge forces of nature are both potentially present within the dynamics of superstrings as they vibrate and move in their ten-dimensional space. In fact, once physics drops thinking about point particle theories in favor of string theories, it becomes possible to think of the whole topic of interactions and forces in an entirely new way.

In the past, physicists had to come up with theories and explanations of how various forces work. Take, for example, the action of a magnet. The properties of a lodestone were known at the time of the Greeks, and in the sixteenth century Queen Elizabeth's physician, William Gilbert, wrote a famous treatise on its properties. But despite these observations, no one really knew how the magnetic force operated. Possibly it was some invisible fluid flowing out of the magnet itself. It was James Clerk Maxwell who proposed that both the magnetic and electrical forces are carried by a field—the electromagnetic field. This idea of a field was a powerful new concept. Less than a century later, after quantum field theory had been created, the electromagnetic field was pictured in terms of quantum particles called photons.

A similar story holds for the weak and the gluon forces. First the physical consequences of these forces were observed and measured, and then physicists had to try to account for their data in terms of gauge fields and the exchange of quantum particles.

When it comes to superstrings, the story is very different. To explain how superstrings interact, there is no need to introduce any forces from outside or to postulate the existence of new quantum particles. Superstrings create their own interactions, which emerge out of the theory in a perfectly natural way. In Green and Schwarz's theory, these interactions are already built in as an inevitable consequence of the topological nature of the strings themselves. Once you assume that nature is created out of extended objects like strings, then interactions and the forces of nature seem to follow.

The basic idea is that strings are free to split and join. An open string can break in two. Two strings can join to form a new string. An open string can join its ends to form a closed loop. Once this is realized, then it is apparent that a host of general interactions can take place between strings. A few possibilities are pictured in the following diagram.

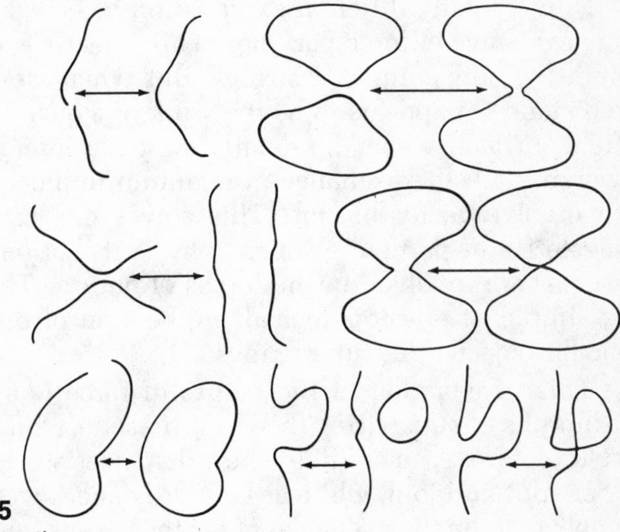

Figure 5–5
String interactions

These pictures have something in common with the Feynman diagrams that were used to calculate the interactions of elementary particles. Although string splittings and joinings do not look at all like the exchanges of elementary particles, in both cases it is possible to represent in a diagrammatic fashion all the possible ways in which two things can interact together. In the case of two strings, they can join and then split apart, or the intermediate string can be allowed to split and rejoin. By adding together all these possibilities, physicists can calculate the overall size of the interaction between strings.

Each string has a label, a series of quantum numbers that identify it. As strings join or split, it looks as if the quantum numbers are changing. From a distance or at low enough energies, where the strings look like points, it will appear as if quantum numbers are being exchanged between point particles. But this is exactly what the conventional point particle theory has always pictured—interactions take place via the exchange of vector bosons like the gluon, photon, and W and Z particles.

Now a totally different explanation has been given for the exchange of force particles. This process is really the splitting and joining of strings. But when viewed from a distance, it appears to be the work of a quantized force field. In fact, when the details of string interaction are worked out, the exchange of quantum numbers appears to involve vector bosons. This solves one of the great mysteries of particle physics: why is it that only vector bosons are involved in the forces of nature? The answer is simple: it is a topological implication of one-dimensional objects, the superstrings.

This explanation of the nature of force is one of the triumphs of superstring theory. It presents a highly intuitive picture. At incredibly short distances we see strings that split and join, while at larger distances we see what appears to be the exchange of vector bosons, the elementary particles of a gauge field that turn local symmetries into global ones.

Strings and Gravity

The interaction picture is even more exciting when it comes to closed strings. To see what happens, it is necessary to go to a full space-time picture. An open string, it will be recalled, traces out a world surface in space-time:

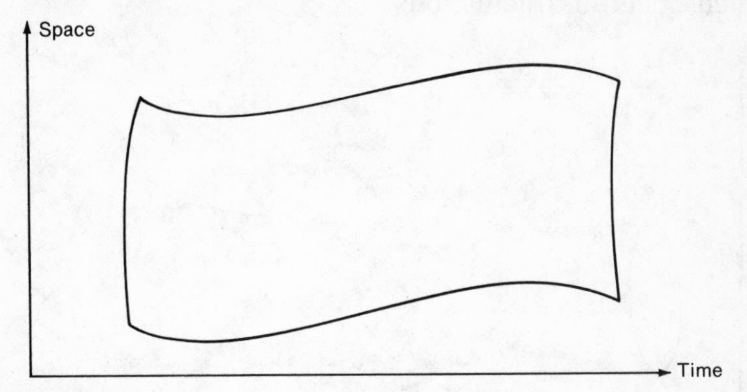

Figure 5–6
The world surface of a superstring. Note that the directions of the time and space axes have been interchanged so time is read from left to right. This change will make the interpretation of string diagrams easier to follow.

A closed string, therefore, traces out a world tube.

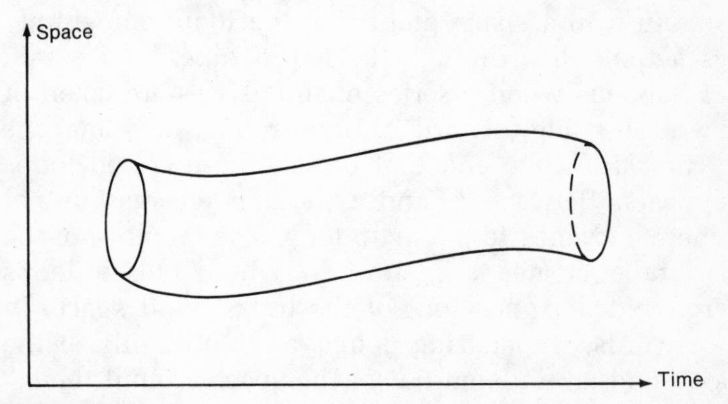

Figure 5–7
The closed world surface of a superstring loop forms a cylinder in space-time.

But this picture is too simple. The quantum fluctuations of the superstring world cause this world tube to fluctuate. The result is similar to a soap film drawn out into a tube that shimmers and blows in the wind. The following diagram shows a picture of one of these fluctuations. Remember that this is only one of a myriad of such quantum fluctuations.

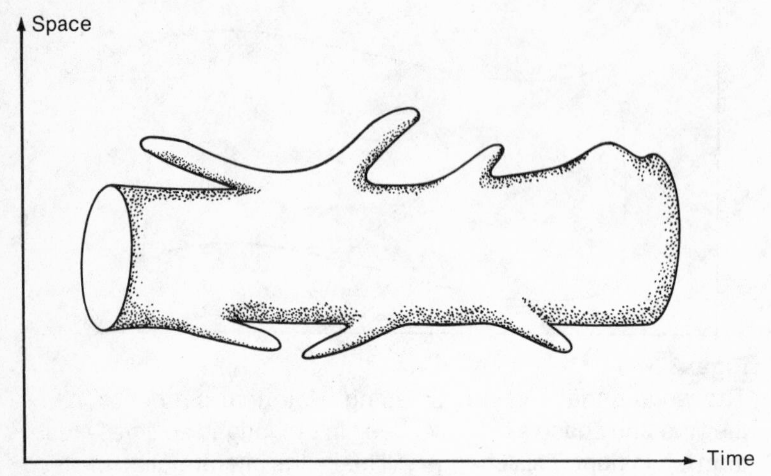

Figure 5–8
Quantum fluctuations change the geometry of the world surface but leave its topology unchanged. It remains topologically equivalent to a cylinder.

To return to a space picture of the loop, one simply takes a time slice through the world tube. Now watch what happens when a series of time slices are taken. It appears, in going forward in time from *A* to *B*, that the loop breaks in two and that one of these closed loops disappears. Between *C* and *D* a loop emerges out of nowhere and joins to the main loop. The topology of the world tube creates a picture in which closed loops emerge and disappear out of the background space. In other words, superstring loops are constantly being created and annihilated out of the space around them.

Superstrings are characterized by their ends, but a closed loop has no ends. At first sight, therefore, it does

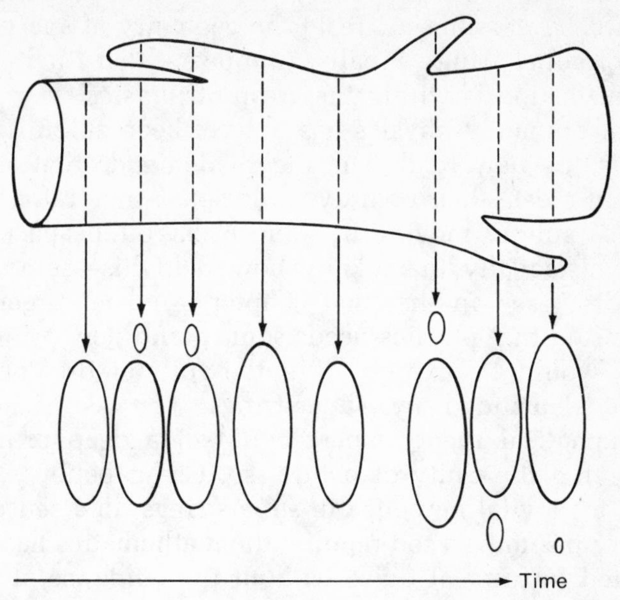

Figure 5–9
Taking successive slices in time, a particular set of quantum fluctuations—distortions of the world surface—can be interpreted as the interaction of closed loops. Note that some of these loops appear to be born out of, and to die back into, the background space.

not seem to have any quantum labels and looks identical to the background space, to the vacuum itself. This agrees with the interpretation that an exchange of closed loops looks just like a spin 2 vector boson, which is hypothesized as the quantum particle of gravity. So the world tube picture shows that gravitons are constantly being born and dying into the quantum vacuum of space. In fact, there is no way of separating the closed loops from this vacuum. Space is no longer a background or backdrop to physics but is intimately tied up with the whole superstring picture.

Space and superstrings must always be taken together. The theory of the elementary particles and of quantum space itself is already unified. Gravity appears on an equal footing with the other forces of nature, and it should be possible to create the geometry of space-time out of superstrings. Finally, it appears, that final dream of unification is within the grasp of physics.

Unfortunately physics has not yet been able to go far along this new road. The theory demands that strings should create their own dynamic space-time rather than strings simply moving in some background space. But physicists really do not know how to do this—some hints are discussed in the final chapter. Michael Green has suggested that physics needs some grand new principle. This would be expressed in the full ten-dimensional space demanded by superstrings. Just as Einstein's mathematical theory emerged out of a deep principle involving the unity of nature, so, Green believes, it is time for a vital new idea in superstrings. In a sense, the theory has moved too rapidly; the mathematics has been pushed forward at times without the guidance of some underlying principle. The real message of superstrings, Green believes, awaits a vital new intuition.

At present, the best physics can do is to throw away the beautiful potential of the closed-loop picture in which space and strings emerge out of the one theory and, in its place, to treat closed-string interactions as if they were happening in some fixed background space. Like a Sleeping Beauty, space-time has yet to awaken, and present-day superstring theory is a pale foretaste of its future possibilities.

This closed-loop picture, when it is fully worked out, will provide a deeper theory of space-time and gravity than Einstein's. At large distances, strings look like points, and their interactions with the background space cannot be seen. But at very small distances, deviations from Einstein's theory will appear. Just as Newtonian gravity became an approximation to Einstein's theory, so too Einstein's picture may one day become an approximation to superstring theory. (There is also the possibility that a new particle, called the dilaton, is predicted by superstring theory. This dilaton, depending on whether it is massless, can also have gravitational effects and may have an important role to play in determining the cosmological constant. It is discussed in the final chapter.)

A further implication of this potential superstring theory of gravity concerns the nature of black holes. Einstein's theory includes the possibility of singularities, points in space at which the whole structure of space-time breaks down. But what will happen in a superstring universe in which points are no longer of primary importance? As a star collapses, its radius becomes smaller and smaller, and in the process, enormous gravitational energy is released. But such energy has the potential for curving space-time—remember that both energy and matter will curve the fabric of space-time. Conventional theory has it that the gravitational attraction will cause the star to collapse right down to a dimensionless point and along the way create a black hole. Einstein's general theory of relativity permits such singular points, at which the fabric of space-time breaks down, and indeed all the laws of physics vanish. Surrounding the space-time singularity is an event horizon—the horizon may be typically 1 km in diameter. Nothing can escape from within this event horizon, not even light itself. In a sense, even light is not moving fast enough to escape from the event horizon of the black hole. Therefore anything that crosses this event horizon is doomed to be swallowed up by the black hole.

But what happens when a collapsing star reaches 10^{-33} cm—the dimensions of the superstring? No one really knows, but some theoreticians have speculated that vibrating superstrings, while not eliminating black holes, could help to avoid the creation of singularities. Shrunk down to 10^{-33} cm, the star will occupy the space of a superstring, and the tremendous energy that is released in gravitational collapse can now be used to excite the vibration and rotation modes of the string. Since there are an infinite number of such modes, they will soak up even the vast energy of a collapsing star. The star need shrink no further. Although it already occupies an unimaginably small distance in space, it will never collapse to a dimensionless point. While the black hole itself will still exist, it need not contain a singularity at its heart.

Many physicists would then breathe a sigh of relief, for this means that the laws of physics need never break down at the point of the singularity. What John Wheeler calls "the crisis in physics" would be averted. (But note that all this does not mean that black holes will not exist, only that they will no longer contain pointlike singularities at which the laws of physics break down. Even though the enormous energy of a collapsing star is used to excite the various vibrations of a superstring, this energy can never be radiated away—light simply cannot move fast enough to escape from the event horizon of a black hole.)

Infinities and Divergences

Now that strings have been allowed to interact by splitting and joining, it is possible to calculate the strength of the various interactions involved. In point particle theories, this is done using Feynman diagrams. Stanley Mandelstam from the University of California and A. M. Poylakov of the Landau Institute for Theoretical Physics in Russia had shown how it was possible to generalize the Feynman picture for the early string theories. Now with Green and Schwarz's approach, it became possible to calculate the various energies of interacting objects.

In conventional point particle theories, the summation of Feynman diagrams often blows up. Interaction terms become infinite as the distances involved become vanishingly small. But in string theory, these short distances are never reached, and anyway very large energies could always be absorbed by the vibrations and rotations of the string itself. Superstring interactions need never blow up.

Moreover, when it comes to the various problems that

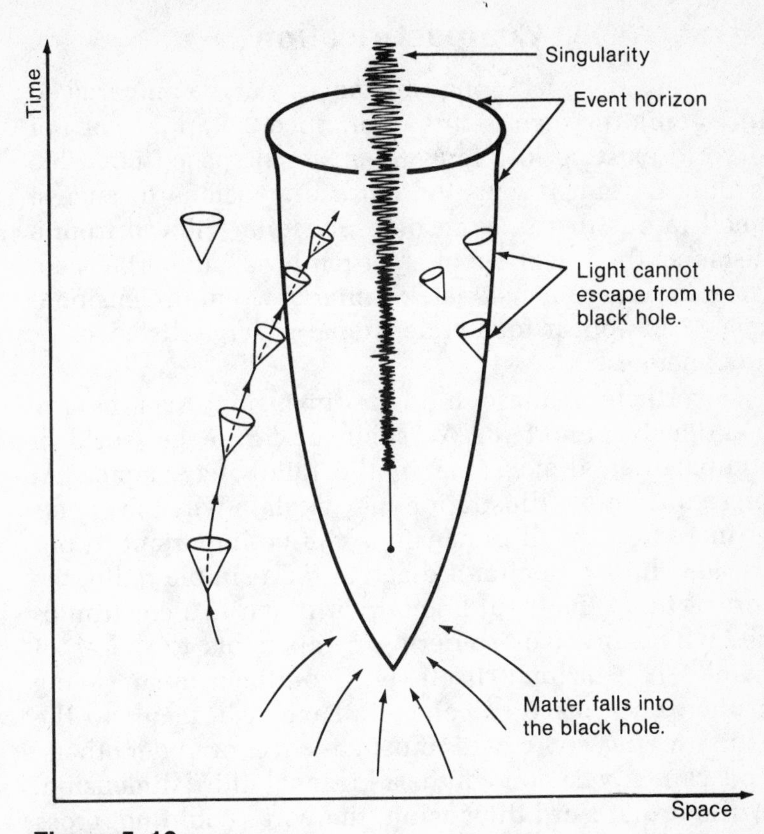

Figure 5–10
A black hole containing a central space-time singularity. Note that the light cones become progressively tilted toward this singularity. At the *event horizon,* these cones are tilted to such an extent that light can never escape. The event horizon represents the point of no return around a black hole.

had plagued calculations in earlier theories, Green and Schwarz discovered that these problems were exactly canceled by the gravitational parts of the theory. Whereas in earlier theories gravity had been an added complication, now it was necessary for the theory to work consistently. Superstrings are not only unified with gravity, they *have* to be unified to make sense.

Compactification

Finally, at large enough distances (large compared to 10^{-33} cm), the theory must reproduce the appearance of a world existing in a four-dimensional space-time. This is done by curling up the extra dimensions to such a small radius that they become invisible. Viewed from a distance, the thickness of a garden hose cannot be seen; it looks one-dimensional. Similarly a ten-dimensional space can appear four-dimensional at large distances or low energies.

Nevertheless, these hidden dimensions can be tremendously important. We shall look into the world of compactified dimensions in the following chapter. At present a simple illustration may explain why the effects of incredibly small dimensions can be important in our world. Think of a circuit diagram for a simple radio, the sort of thing that could come with a child's electronics kit, which involves soldering components to pieces of wire. The diagram itself is two-dimensional, being printed on a piece of paper. But when it comes to the actual wiring, some wires must pass over or under others. This can only be done in the additional, third dimension. Without that third dimension, the wires could not cross each other without touching.

Now imagine a giant piece of electronics that stretches across a football field. To all appearances it would be two-dimensional; the only property of the third dimension that would ever be used would be the few millimeters thickness of two wires. From a distance the third dimension of this circuit would be invisible. Nevertheless the existence of the hidden third dimension makes it possible for complex interconnections to be made without the wires actually touching.

So hidden dimensions can be of vital importance. In fact, in superstring theory it is the six compactified dimensions that, in a sense, make the whole theory work. Without them superstrings would never be able to exhibit the full symmetries of nature.

In particular Green and Schwarz's theory was based on a unifying symmetry called SO(32). These symmetries will manifest themselves at the high energies and short distances in which physics is unified. But at lower energies and longer distances, this symmetry is broken and we see nature in terms of various gauge forces and point particles with different masses. It is important, however, that as the six dimensions curl up, the breaking of the initial grand symmetry should proceed in exactly the right way to reproduce all the symmetries and behavior of our four-dimensional world.

But does compactification really work? Is there a coherent account for the mechanisms by which our four-dimensional universe is born out of a world of ten dimensions? This is still a matter of great controversy; discussion will be postponed to the following chapter and to Chapter 10.

Conclusion

In summary, Green and Schwarz had finally put all the pieces of the puzzle together and had satisfied the enormous constraints imposed by chirality, grand unified symmetries, relativistic covariance, and quantization. The result was a triumph, a theory free from all the drawbacks, anomalies, and infinities that had been dogging particle physics for decades. In addition, this theory also produced new insights into the nature of force and the connection between gravity, space, and the quantum theory. But physics does not remain stationary, and not long after this theory of superstrings had been published, theoreticians were exploring new ideas, in particular the deeper meaning of closed loops.

6
Heterotic Strings: Two Dimensions in One!

THE PUBLICATION OF Green and Schwarz's "Anomaly Cancellations in Supersymmetric D = 10 Gauge Theory and Superstring Theory" in 1984 had an effect similar to that of dropping an opened jar of honey near an ants' nest. Soon an army of theoretical physicists was swarming around the theory. Indeed it looked as if there would be room for everyone in this world of superstrings.

The historian of science Thomas Kuhn has pointed out that the power of a new scientific theory lies in its ability to create work for the professional community. In this respect, Green and Schwarz had hit the bull's-eye, for theoreticians of all shapes and sizes were bending their talents to this new theory. They examined the arguments used by Green and Schwarz, checked their calculations, looked closely at the symmetry groups, confirmed that interactions did not blow up and anomalies were absent, rewrote parts of the theory in different ways, and made connections to other ideas. A regiment of "phenomenologists" (physicists who attempt to make connections between the grand and austere predictions of new theories and actual experimental results) worked out the implications of Green and Schwarz's paper. They looked for ways of breaking down the symmetry of the group SO(32) demanded by the open superstrings to the point where they could make connections with experimental data. Then there were the mathematical physicists who explored the formal implications of superstrings, refined the arguments, and attempted to generalize some of the mathematics used in the theory. There was work for everyone, from the run-of-the-mill particle physicist looking for a thesis problem for a student to the theoretical giants.

One of the groups who focused their attention on superstrings could be found at Princeton. David Gross, Jeffrey Harvey, Emil Martinec, and Ryan Rohm became known as the "Princeton String Quartet." This group was studying Green's and Schwarz's early approach to Type II, or closed-loop, superstrings. While Green and Schwarz had been able to show that their closed-loop theory was free from infinities, they had eventually dropped the idea of closed strings because they did not seem able to explain the gauge forces of nature. In their place they concentrated on their earlier, Type I approach, and by 1984 had shown that such a theory can accommodate nature's forces, is chiral, and is free from infinities and anomalies. Nevertheless, Gross and his colleagues wanted to go back to the closed-loop theory, but this time writing it down in a new way.

In Green and Schwarz's Type I approach, the breaking and joining of open strings had been equated with interactions, and quantum numbers had been located on the string ends. An exchange of closed loops had therefore been identified with interactions of the background space or quantum vacuum. Since loops have no free ends on which to carry quantum numbers, they appear identical with the vacuum of space. So a theory that only includes closed loops did not look as if it could account for the nuclear and electroweak interactions.

But there are other ways of thinking about closed-loop theories. For example, it is possible for fields to actually run around the strings themselves. In Green and Schwarz's Type II theory, the two physicists had allowed fermion fields to run in one direction while boson fields ran in the other direction. But this means that it is possible to have quantum numbers, the labels we associate with quantum states, run around the loops themselves. Rather than associating quantum numbers with ends of open strings, they can actually rotate around the closed strings themselves.

The idea is a little like that used to store information in a computer. Certain types of computers use magnetic rings to store bytes of information. This information circulates around the ring as binary pulses. In an analogous way the coordinates used to label fermions and bosons now run around a closed-loop superstring.

When the mathematics is worked out, it appears that a closed-loop superstring will allow quantum numbers to travel both clockwise and counterclockwise. In fact, it is possible to have two waves of quantum numbers traveling in opposite directions at once without ever getting them mixed! This means that the boson quantum numbers can circulate in one direction while the fermion numbers circulate in the other. Bosons and fermions become unified while, at the same time, remaining separate.

But Gross wanted to go much further. In essence he wanted to combine what look like two very different theoretical approaches within a single formalism. Remember that Nambu's original bosonic string theory had required a space of twenty-six dimensions. On the other hand, the original fermionic string theories demanded only ten dimensions of space. But would it be possible, Gross speculated, to create a hybrid of these two approaches, a theory that is both ten- and twenty-six-dimensional at the same time? A string with two different dimensions associated with it is heterotic. The word *hetero* comes from the Greek and implies the combining of two or more different things—in this case different spaces.

To put Gross's proposal more formally, the fields that describe the physical degrees of freedom of the string in its ten-dimensional universe can be divided or decomposed into two independent parts. One part moves clockwise while the other circulates in a counterclockwise direction around the string. It turns out that the description of the boson field requires sixteen extra coordinates. Later, when the ten-dimensional space in which the string moves is compactified, it will turn out that these sixteen boson dimensions have all the necessary richness to account for the gluon and electroweak gauge fields. Interactions are therefore hidden within the string as it moves in its ten-dimensional space.

It is at this point that the idea of dimensionality begins to boggle the mind. Just what are the dimensions of this heterotic string theory: two, four, ten, sixteen, or twenty-six? In one sense, the two-dimensional world sheet or world tube of the string itself could be thought of as the essential dimensionality. On the other hand, these strings are said to move in a ten-dimensional world, which in turn compactifies down to our own four-dimensional space-time. But now David Gross is suggesting that within the string itself there is a sixteen-dimensional bosonic field!

Things are becoming even more complicated today, for to jump the gun a little and anticipate some of the results of the final chapter, some physicists are arguing that these "extra dimensions" may not be dimensions at all!

In Gross's original version, a heterotic string combines a ten-dimensional fermionic field moving to the right with a twenty-six-dimensional field moving to the left. More recently physicists are speculating that four dimensions may be the natural way of describing space-time. But this means that the heterotic string could be thought of as referring to four dimensions with, moving to the right, six additional "quasi dimensions" (4 + 6 = 10), things that are not dimensions in the real sense of the word. Likewise, moving to the left, there will be twenty-two "extra dimensions" (4 + 22 = 26). But exactly what these extra six and twenty-two quasidimensions may be is, at the moment, something of a mystery. Suffice it to say that when Gross first proposed his heterotic theory, he had in mind combining ten and twenty-six dimensions within a single theory.

This then was the essential idea proposed by Gross and his co-workers: that closed heterotic superstrings, as they are called, carry all the quantum numbers they need around their loops. Gross discovered that it was indeed possible to create a consistent theory in this way, and one that would not include infinities. As with Green and Schwarz's approach, the key lay in choosing the correct overall symmetry for the theory. This time it turned out to be a choice between SO(32) and E8 × E8.

Gross also invoked a sort of stringy version of the Kaluza-Klein theory in which the forces of nature emerge as the space of the heterotic strings compactifies. While Gross's theory offers two choices of symmetry, SO(32) and E8 × E8, it is the latter symmetry that appears more attractive to the elementary particle physicists. The exact meaning of this symmetry need not concern us here; the basic point is that this E8 × E8 symmetry exists in the full ten-dimensional space in which the string moves and vibrates.

But what happens when this space begins to curl up or compactify? The result is that the basic symmetry pattern E8 × E8 begins to break. The fundamental assumption of heterotic string theory is that, through a series of symmetry breakings, the initial grand symmetry pattern, E8 × E8, will eventually be reduced to the more familiar symmetries that are characteristic of the known elementary particles. If this happens in exactly the right way, then it holds out the hope that the elementary particles can somehow be recovered, in the low-energy limit, from string theory.

First let us see what happens to the initial grand symmetry, E8 × E8. Some physicists hypothesize that this corresponds to two universes, each corresponding to the smaller symmetry pattern E8 by itself. In addition to our own universe, there is a new, hypothetical universe, a shadow world as it were. Combining these two universes—each with its own E8 symmetry—produces a grand E8 × E8 symmetry. Or at least this is one way of trying to explain how the initial heterotic symmetry is broken down.

The basic idea is that each E8 group describes its own universe, its own pattern of particles and forces. But now something curious happens. Each of the two groups is fully able to account for the known elementary particles and forces of nature. In other words, each group is complete unto itself. Moreover, since the forces of nature, other than the gravitational force, are confined within each group, the elementary particles in one group are effectively invisible, or hidden, when viewed from the other group. It is as if when E8 × E8 breaks, it does so into two universes, one like our own and the other an invisible shadow world. E8 × E8 therefore becomes E8 and E8(shadow).

But let us forget about this hypothetical shadow universe for a moment and give our attention to the E8 symmetry of our own world. Gross postulated that the symmetry of the remaining E8 group is further broken. The flat ten-dimensional space in which the heterotic strings vibrate and move begins to curl up—or at least six of its dimensions curl—leaving our familiar four-dimensional space-time. The first stage of this symmetry breaking is to produce the smaller group E6. This further breaks down into SU(3) × SU(2) × U(1). But this is exactly the symmetry demanded by the grand unified theory. Remember that SU(3) is the symmetry of the quark theory or standard model, while SU(2) × U(1) is the symmetry of the electroweak interaction—the unified form of electromagnetism and the weak nuclear force. Combining these symmetries produces SU(3) × SU(2) × U(1), and this very symmetry appears to be generated out of the basic superstring symmetry as the ten-dimensional space curls or compactifies.

Witten and his colleagues have pointed out that a consequence of this symmetry breaking is that the elementary particles seem to replicate themselves, possibly to form a quartet of families. For example, each family or pattern of particles is reproduced several times with the families being identical in every respect except for differences in their masses.

This seems a curious result, that nature would involve itself in such replication. However, the known leptons themselves do in fact exhibit a redundancy: In addition to the electron there is also the tau and muon, identical to the electron in all respects except for their masses. Furthermore, each of these particles has an associated neutrino. The leptons therefore appear to occur in triplicate. Is it possible that a further partner exists to form the quartet suggested by Witten?

Compactification of the original ten-dimensional space in which the string is defined progressively breaks the heterotic string symmetry down to the point where we begin to see the hadrons and leptons of more conventional theories. In addition, this symmetry breaking allows the particles of the gauge fields, the gluon and electroweak vector bosons, to manifest themselves. Viewed from a distance, the symmetry-broken heterotic strings look just like familiar point particles—but without the infinities and anomalies of the particle approach.

Yet, in a deeper sense, the theory is ambiguous, for a full description of the compactification process is missing. Indeed it is theoretically possible for E8 to break down in a number of different ways, some of which will produce the symmetry schemes that were used to group particles together in the earlier grand unified theories.

The accepted idea is that E8 breaks to $SU(3) \times SU(2) \times U(1)$, but the mechanism for this breaking remains to be justified theoretically. The hope is that when a theory of the compactification process is finally developed, it will indicate the precise steps by which the original heterotic symmetry breaks down. It should also determine the exact symmetry patterns of the elementary particles and their individual masses. All that can be done at present is to point out that E8 had indeed been chosen previously by grand unified theoreticians as a plausible unifying symmetry in their attempts to explain the underlying order of particles and forces.

Interactions

The heterotic string consists only of closed loops, so there can be no interactions involving the ends of a superstring, as happens with Green and Schwarz's theory. In fact all the myriad processes of nature are ultimately reduced to the amoeba-like budding and rejoining of loops. To follow the interactions of heterotic strings, we must return to the world tube illustrated in the previous chapter.

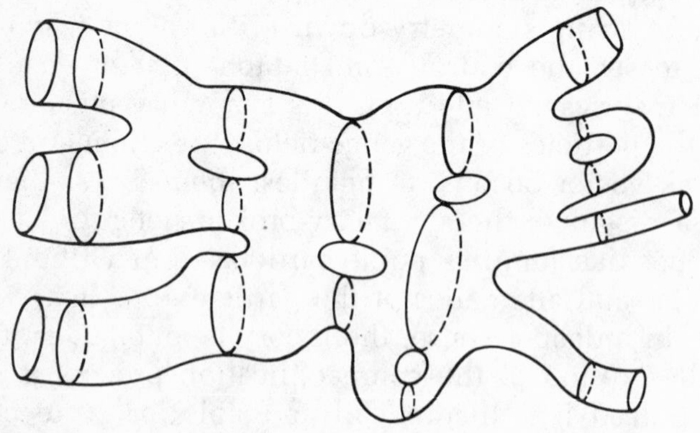

Figure 6–1
The world surface of a heterotic superstring with its quantum fluctuations

Since all superstring interactions must ultimately come from such tubes, it is worth taking a closer look at what exactly is happening. Look, for example, at all the possible ways in which two free loops can come together and interact. The first possibility is that they approach, merge, then break apart again.

Figure 6–2
Two closed strings interact by joining and separating.

Take a segment of the related world tube. It looks a bit like two pairs of trousers that are joined at the waistband.

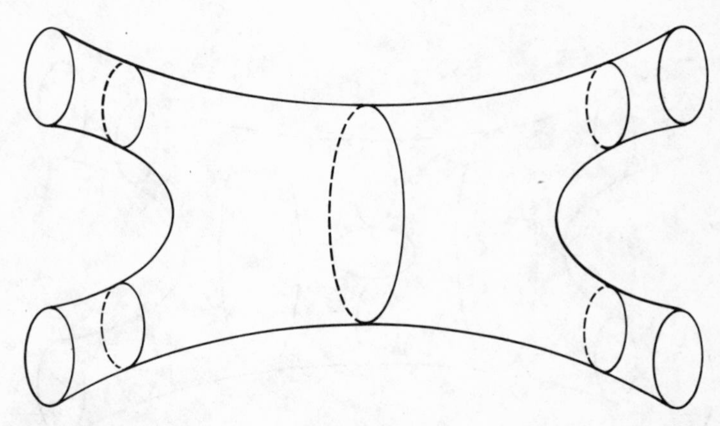

Figure 6–3
Drawn out in space-time, the joining and splitting of a pair of closed-loop strings has the appearance of two pairs of trousers joined at the waistband.

At the next level of interaction, the two strings merge, break apart, merge again, then finally go their separate ways.

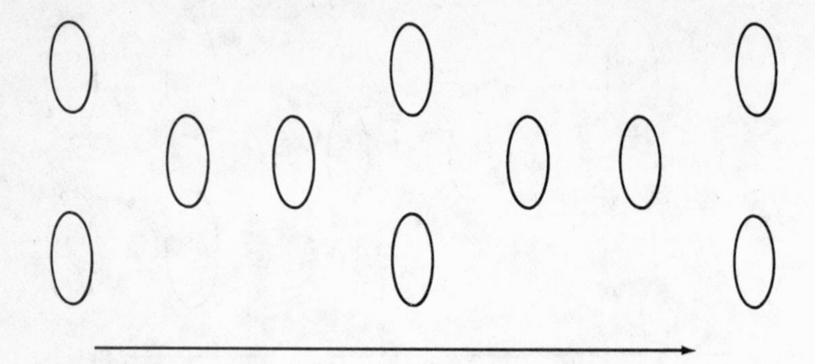

Figure 6–4
A pair of strings join together, pass through an intermediate stage in which they split, join again, and then separate.

This time there is a hole in the center of the trousers.

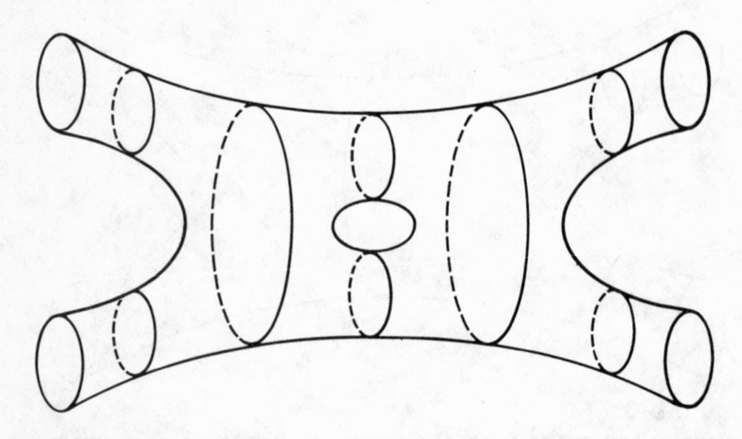

Figure 6–5
The more complex string interaction of Figure 6–4 is represented in space-time by a trouser diagram having a hole in the waistband.

As you will anticipate, the next stage in complication will involve a pair of trousers with two holes in the center.

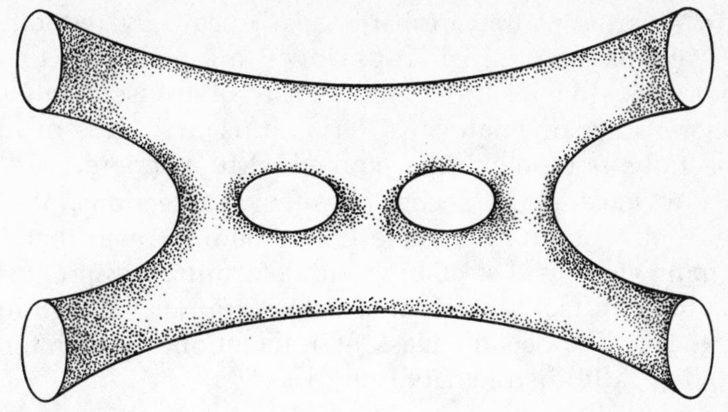

Figure 6–6
A trouser diagram containing two holes. In general such a diagram can have any number of holes in the waist and will correspond to multiple interactions of closed loops.

Now let's pull out the mathematical stops and generalize the whole process. According to the theory of superstrings, stretching and contracting the trousers in a diagram like the following one has no effect upon the physical value of the interaction. (Remember how, in the previous chapter, it was possible to relabel the coordinates within the world sheet without changing the value of the action.) In other words, it is possible to carry out a topological transformation upon the world tube, to stretch and deform it, provided we do not leave behind any rips or tears at the end of our manipulations.

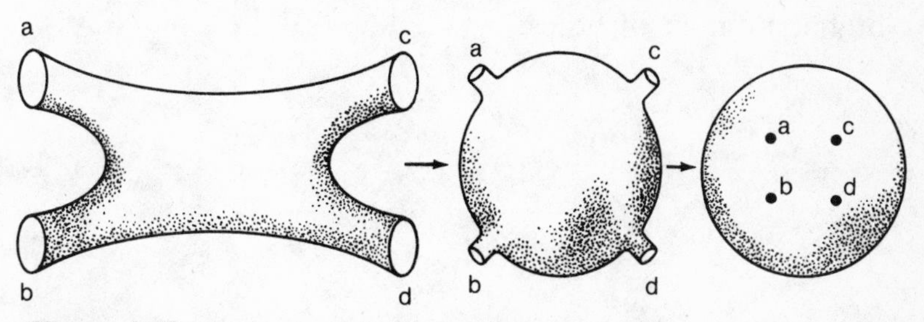

Figure 6–7
The trousers can be deformed without changing their underlying topology.

In a series of transformations, we begin by reducing the size of the original loops down to a point and then deforming the resulting linked pair of trousers into a sphere. In mathematicians' terms, this first class of interactions is topologically equivalent to a sphere.

Now look at the second level of interactions. With trousers containing a single hole where the waistbands join, no amount of stretching and deforming will produce a sphere. In fact, the result of deformation is a doughnut or torus. The second class of interactions is therefore topologically distinct from the first.

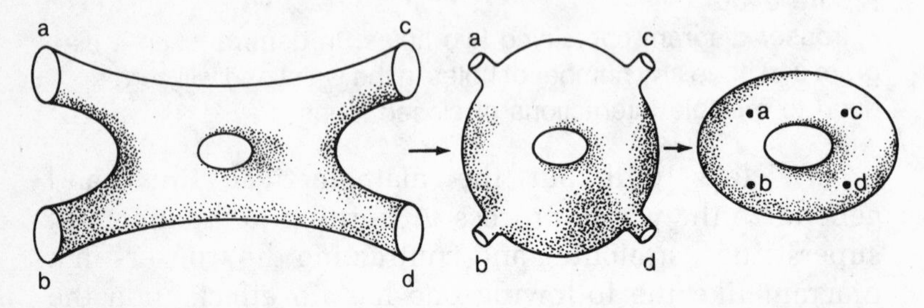

Figure 6–8
Trousers with a single hole in their waistband are topologically equivalent to a doughnut. These sugar-coated trousers have yet to catch on at fast-food outlets!

With the next class we get a torus with two holes. It is easy to see that the hierarchy of possible interactions goes from sphere to torus to torus with a higher and higher number of holes.

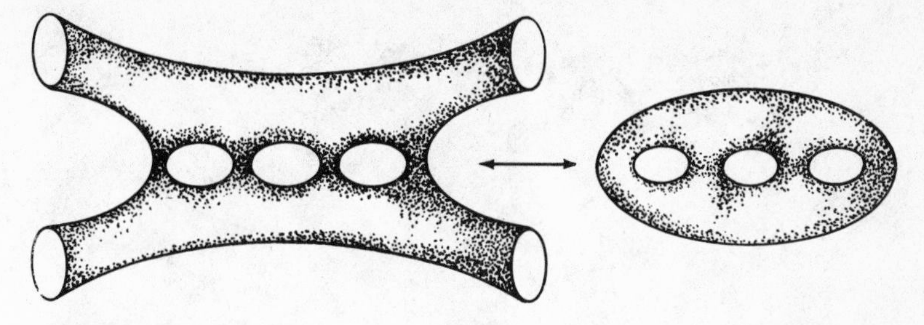

Figure 6–9
Trousers with *n* holes in the waistband are topologically equivalent to an *n*-torus, a doughnut with *n* holes.

This result, which is mathematically attractive, completely describes the interactions of heterotic superstrings in topological terms. The possible trouser diagrams that can occur when strings interact are incredibly simple when compared to the complexity of the Feynman diagrams in a world of point particles. Superstrings have simplified the way physics is able to deal with interactions.

It is possible to go further. Topology tells us that if we cut the torus, give it a complete twist, and then join it up again, nothing significant will have changed. In other words, heterotic string theory must be symmetric under such a topological transformation of the interactions. But what overall symmetry is unchanged by such a topological transformation? Hold your breath—it is E8 × E8! The basic unifying symmetry appears again, the same symmetry that was earlier demanded by the requirements of chirality.

Interactions and Infinities

Yet more has to be gleaned from the interaction diagrams. Take the trousers in the following diagram.

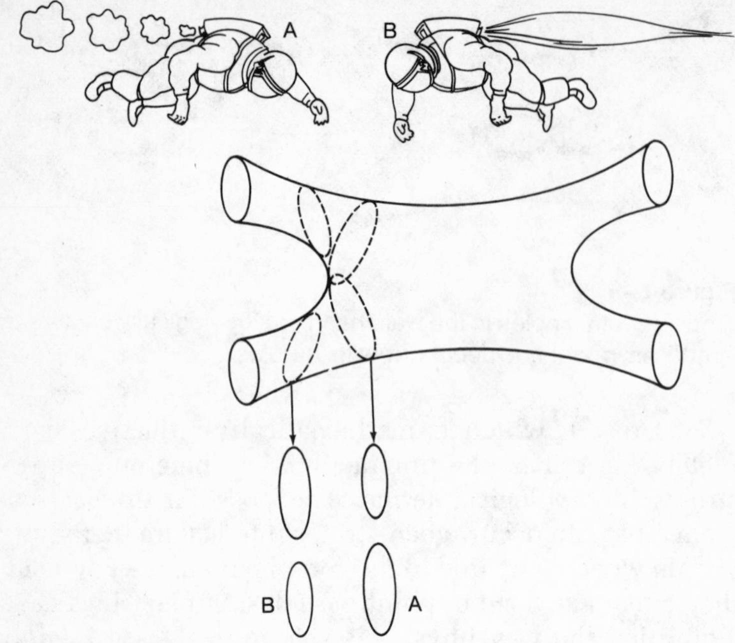

Figure 6–10
The way in which a space-time slice is taken through the world tube of a heterotic string depends upon the observer's state of motion. While one observer may see the two loops touch at point *A*, an observer moving at a different speed sees them touch at point *B*.

Suppose an observer *A* takes a close look at this interaction. It appears that two loops are just touching, so the distance between them is zero. The infinities that traditionally plagued point particle physics can generally be traced to the fact that interactions get bigger and bigger as interacting particles approach. When their separation becomes zero, the strength of their interaction becomes infinite. But what about the superstrings seen by observer *A*? Their separation looks as if it is zero. Does this mean that an infinite interaction will result?

But there are other observers, *A′*, for example. This observer moves at a different speed and therefore takes a different view, a different slice through space-time.

Whereas *A* sees two loops in contact, *A′* sees two loops separated by a finite distance. Events, as Einstein pointed out, appear different to different observers. In fact, there is an infinity of possible observers, each moving at a different speed through space-time and each seeing the heterotic strings at a different separation. In this way, relativity makes it possible to avoid infinities in the heterotic string theory.

The Shadow Universe

To make physical sense of Gross's heterotic string theory, it is necessary for the basic symmetry, E8 × E8, to break in two. E8 had earlier been chosen by a number of grand unified theorists as the starting point for their theories of point particles, for it is more than large enough to accommodate all the particles and forces of nature. In other words, breaking the initial heterotic symmetry appears to create two identical universes, E8 and E8 (shadow). We live in our universe with its planets, stars, and galaxies, but according to this idea, there could exist another, parallel universe, invisible to us but in general respects identical!

Each world has its electromagnetism, weak and strong interactions, protons, electrons, and mesons. These elementary particles can build themselves into atoms, and these atoms into molecules. Finally molecules form rocks, planets, stars, and even living systems such as theoretical physicists and the readers of science books like this one. Yet the electromagnetic and nuclear forces of one world will have no effect in the other. It is as if E8(shadow) refers to an invisible universe, another world that coexists and interpenetrates our own. The electrons of the shadow world cannot interact electromagnetically with our electrons. Protons and shadow protons do not see each other; shadow rocks pass right through our walls. Since the gluon and electroweak forces are confined within each of the two E8 groups, they cannot give rise to any form of interaction between our own and the shadow universe. Yet both universes exist within the one space-time, and since gravity is a property of the full E8 × E8 group, it will survive this symmetry breaking and compactification. Gravity links these two universes; they coexist within a single space-time that is curved by their combined mass and energy.

We know, of course, that our own universe exists. But is this shadow world a fiction, a figment of a theoretical physicist's imagination? Close analysis indicates that it is indeed theoretically possible for a shadow universe to exist in parallel to our own. While we would feel its gravitational effects—since gravity is a property of the full E8 × E8 group—this shadow universe would be otherwise invisible. Photons from the shadow group E8 have no interaction with the matter in our universe. As you are reading this book, you could be occupying the same chair as a reader from the shadow world. Or you may be sitting at the bottom of a shadow ocean, or at the heart of a shadow sun. For once science has outstripped science fiction.

This idea of a shadow universe was not unwelcome in certain areas of physics, for it offered a solution to the *problem of missing mass*. The general theory of relativity makes it possible to say something about the size, rate of expansion, and degree of curvature of the universe in relation to the amount of matter and energy it contains. Current theories of the large-scale structure of the universe use the results of astronomical measurements to predict its total mass.

But the problem with this theory is that the predicted mass is larger than what is observed. When the mass in all the stars and intergalactic dust is added together, it turns out to be too small. Either these theories of the structure of the universe are wrong, and this would probably mean that general relativity is also wrong, or some additional mass must be hidden from us.

The idea of a shadow universe is an ideal way of explaining missing mass. The shadow universe is invisible; it cannot be seen or touched, yet because gravitational forces make no distinction of this division between the two worlds, its mass will affect the overall structure of this one space-time.

Not every physicist takes the idea of a shadow universe that seriously. But Edward W. Kolbe, along with David Seckel and Michael S. Turner of the Fermilab in Illinois have worked out the history of a hypothetical shadow universe. The story begins at the big bang, the moment of creation. In that first instant, all physics is unified, a single universe exists in a ten-dimensional space in which all dimensions are no bigger than the size of a superstring. The symmetry of this embryo universe is given by the full E8 × E8 group. But now space-time enters a phase of rapid expansion, or rather three of the space dimensions unfold, while the others are tightly curled. Then the E8 × E8 symmetry is broken in half, and from now on out, the matter that will eventually give birth to our solar system, and to ourselves, is separated from matter in the shadow universe. Yet space expands for both these universes alike, for they both exist in the same space-time.

The shadow universe initially interpenetrates our own, and there is no reason to believe that this does not continue to happen. E8 in both universes undergoes a further symmetry breaking, to produce the grand unified pattern of hadrons and leptons, fermions and bosons. Atoms come together, dust particles coalesce, stars are created, galaxies and planetary systems form. But since these two worlds can sense each other only through their gravitational attractions, at the atomic level the shadow world remains totally invisible. As we drive to work in the morning, our car could be passing through the walls, buildings, and mountains of a shadow city.

It could turn out that a star in our universe is located close to an invisible, shadow companion. Since gravity transcends the breaking of E8 × E8, while the shadow star remains invisible, its attraction will affect the orbit of the normal star. The two will form a binary system in which a star in our universe is rotating around an invisible companion. Such an effect could be detected by conventional astronomical means.

In fact, gravitational effects like these enable us to rule out the more romantic speculations that were made a few paragraphs earlier about shadow cities and shadow readers. The dynamics of our solar system have been worked out to such an accuracy that there is no room for the effects of an interpenetrating solar system with all its additional mass.

But what about a shadow meteor rushing toward the earth? Not encountering any resistance as it reaches the earth's crust, it would rush through our planet like a car though a snowstorm. However, our earth would still exert a gravitational attraction for such a meteor, and it is possible that, given exactly the right conditions, it would eventually wind up at rest at the center of our planet. Could shadow meteors and intergalactic dust reside at our earth's core? Their effect would be to make the earth slightly heavier than we expect from our knowledge of its size, composition, and density. The best calculations suggest that our earth could contain as much as 10 percent shadow matter, but with more than this, the orbits of artificial satellites, which are very accurately known, would be noticeably affected by an increased gravitational attraction. Of course, this does not rule out shadow stars and shadow systems that are only a few light years away from our own, or the possibility that we could at this moment be passing through vast gas clouds of shadow matter.

There is also the possibility that this shadow universe is not in fact identical to our own. Following the first instants of creation, a sudden expansion of the three spatial dimensions occurs, with the other six remaining tightly curled up. At this point, $E8 \times E8$ breaks down into two E8 symmetry groups and two identical universes. In our universe we know that E8 must next break into SU(3) × SU(2) × U(1), and that SU(3) must then separate so that the gluon force can break away from the electroweak force and hadrons can differentiate their masses from leptons. Then SU(2) × U(1) breaks, and the electroweak force becomes a short-ranged weak force leaving the infinite-ranged electromagnetic force behind.

But who is to say that exactly the same symmetry-breaking steps take place in the mirror world? It is possible that E8(shadow) could break in an entirely different way. The shadow universe could consist of elementary particles with entirely different masses, interacting by forces of quite different strengths. The nuclei within this universe would have different stabilities, and radioactive elements would decay at different rates. Since so many nuclear processes are finely tuned, this shadow universe would look very different from our own.

Particle Masses

In going from Nambu's original string theory by setting the lengths of superstrings and heterotic strings so incredibly small, and their tensions so astronomically large, physicists had also thrown away one of the most attractive aspects of the theory, the ability to explain the masses of the hadrons and their resonances. Now the vibrations and rotations of a superstring correspond to a spectrum of masses with enormous separations. While the mass corresponding to the bottom note of the heterotic string or superstring scale is zero, the next note has a mass that is almost comparable with a minute dust particle—absurdly large for an elementary particle. While these highly energetic vibrations and rotations of the string are important in the first instants of the creation of the universe and also help in keeping infinities and divergences at bay, they can have nothing to do with the spectrum of masses of the known elementary particles. The hadronic baby has been thrown out with the bathwater.

In fact, viewed from a distance, so that the superstrings and heterotic strings look like points, all elementary particles have zero mass. Their real masses must therefore be brought about by various symmetry breakings that take place in our actual world. The idea is that, at the very high temperatures during the creation of the universe, the theory would be totally symmetric with all particles in the lowest string state having zero mass—while the masses for excited states would be extremely large. As the universe cooled, however, this initial symmetry was broken. But this means that it is possible to have overall solutions for the whole universe that have a lower energy. It is in this symmetry-broken, lower-energy solution that the particles finally acquire their masses.

However, at the moment, the fine details of this symmetry breaking and the precise values of the particle masses cannot be given by superstring theory. Indeed, the actual masses of the elementary particles would appear as a very fine correction to the basic superstring theory. Clearly what is needed is some new insight, some deeper principle that would explain the mechanisms of symmetry breaking and allow particle masses to be calculated accurately.

What Are the Dimensions of Space-Time?

One of the major challenges presented by superstring theories is to explain the dimensionality of the space-time we live in. We certainly appear to live in a world of three spatial dimensions, but does this follow inevitably from some deep law of nature? It is not too difficult to show that our space cannot have any fewer than three dimensions. Remember the example of the electronic wiring diagram introduced near the end of the previous chapter? This showed that it is impossible to interconnect a large number of elements within a two-dimensional space without having these various interconnections touch rather than pass over or under each other. The complex circuitry of a radio or the nervous system would be impossible in a two-dimensional world, as would be a host of other processes and properties that depend upon the topological interconnectedness. This certainly puts a lower limit upon the dimensionality of space-time, but is there also an upper limit?

Superstrings set this limit at ten dimensions, yet clearly our large-scale world is not of such a high dimensionality. The explanation accepted by most physicists is that six of these dimensions are hidden, curled up with a radius of some 10^{-33} cm. But such an assumption runs into serious difficulties, for it also assumes that these dimensions must have remained rolled up and with exactly the same radius for as long as the universe has existed—leaving out the first unimaginably short instances of its creation.

Yet the space-time in our large-scale world is by no means static. The universe as a whole is expanding, and, on a smaller scale, the geometry of space-time constantly responds to the movements of matter at all levels. Why then, if four dimensions of space-time are in a constant state of flux, should we assume that the other six of them are absolutely stable? Not only does this stability of the six compactified dimensions seem unreasonable on relativistic grounds but on quantum considerations as well. Why does the quantum theory not introduce fluctuations into the radius of these compactified dimensions?

It is also possible that the original ten-dimensional universe could have compactified in some other way with, for example, four instead of six dimensions becoming tightly curved. In such a universe, there would be five large-scale dimensions of space and one of time. Of course, this bears no relationship to our universe. But suppose such an alternative space-time could exist; then it becomes theoretically possible that it could tunnel, quantum mechanically, into our own. While there could be immense barriers of energy that prevent these two universes having any normal commerce with each other, nevertheless there would be a small, yet finite probability that a form of quantum mechanical tunneling could occur from one to the other. The effect would look like a spontaneous uncurling of two of the compactified dimensions within our own universe.

A full superstring theory must explain how space-time obtained its present structure following its initial generation in the big bang. After the first instant of creation of a ten-dimensional universe, it would appear that six of these dimensions became tightly curled while the remaining four engaged in a phase of rapid expansion. Superstring theory must therefore say something about the large-scale structure of these four space-time dimensions. One of the great unsolved problems of modern physics is why the empty space-time of the universe is so flat—that is, why the cosmological constant is zero. Of course, the presence of the sun and planets creates a curvature in space-time, but we also know that in the absence of matter, the space-time is totally flat, to a degree of accuracy as high as one part in 10^{120}. It is possible that one day superstring theory, or its extension, will be able to explain this fact.

Admitting that there are serious difficulties connected with this whole problem of compactification, let us assume that curling up six dimensions of a ten-dimensional space is a *fait accompli* and go on to examine its implications. Heterotic string theory does not exhibit the gluon and electroweak forces in its original ten-dimensional space; they appear only upon compactification. Another way of describing this is that the average value of the fields that carry these forces is zero in a ten-dimensional universe. In other words, the universe lies at the bottom of a potential well, like a roller coaster in a valley.

Figure 6–11
The roller coaster at the bottom of a valley represents a universe in its lowest energy state.

The valley is symmetric at this point, but suppose that its geometry becomes distorted; it is now possible for the roller coaster to move to one side or the other and fall to a state of lower energy. In other words, by falling to a lower energy level, the universe as a whole can break the basic symmetry inherent in its original state.

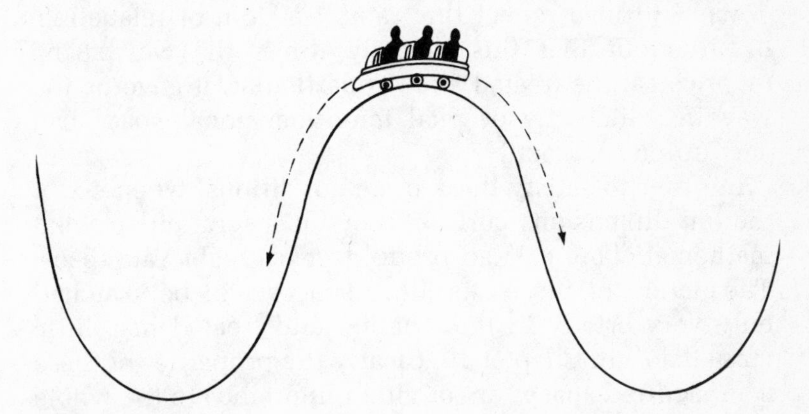

Figure 6–12
A universe could exist in an unstable state, poised upon a hill between two energy valleys. By choosing to fall into one valley and release its surplus energy, the universe will also break a basic symmetry. (Clearly the case where the roller coaster lies in the left or the right valley is less symmetrical than when it is midway between them.)

This symmetry breaking occurs as six of the ten flat dimensions begin to curl. Its result is that the average values of the force-carrying fields is no longer zero. The gluon and electroweak forces appear at the expense of breaking the symmetry of the original universe, that is, of singling out six of its dimensions to become tightly curled.

Orbifolds and Calabi-Yau Spaces

The properties of superstrings arise from their two-dimensional world sheets or world tubes—remember how the one-dimensional string is stretched out in time to create a two-dimensional surface. Now a simple relabeling or interchange of coordinates on this sheet can have no physical effect on the string itself. After all, nothing has really happened, simply a change in the way the world sheet is labeled. This means that physicists must be very careful to ensure that the theories they write down will also reflect this same freedom of relabeling. It turns out that this severely limits the way string theories can be treated and, in particular, it governs the way in which the original ten-dimensional space can curl up or compactify.

In order to satisfy these basic conditions, when six of the ten dimensions curl up, they must form either what mathematicians call an orbifold or a Calabi-Yau space. The nature of these curled-up spaces will be sketched only very briefly in this chapter and treated in a little more detail in Chapter 10. Clearly the properties of these compactified spaces are of vital importance to the whole string theory. At present they are being studied by a number of groups, including teams at the University of Texas and at the University of California, Santa Barbara, as well as by Edward Witten and his colleagues at Princeton.

One important consequence of dealing with orbifolds or Calabi-Yau spaces is that, as the original ten dimensions of string space are curled up, the basic supersymmetry of the theory is not destroyed. This is a key to the whole string approach, for it is not good creating a supersymmetric theory if its vital symmetry is destroyed as soon as the space is curled up.

In addition, the act of compactification will break the original E8 × E8 symmetry of the heterotic strings. It is vitally important that the breaking of this symmetry should take place in exactly the right way. Since the symmetry SU(3) × SU(2) × U(1) crops up in the best accounts of elementary particles, it would be a good idea if E8 × E8 reduced to this symmetry as the six dimensions curled up.

Of particular importance for the six compactified dimensions is their topology. Topologies, remember, are more general than geometries, for they can deal with spaces that can be stretched and distorted and contain a variety of holes and handles. Particular topologies can be characterized by what are called their Euler numbers or Euler characteristics—these are named after the great eighteenth-century mathematician Leonhard Euler. Euler characteristics are related to the dimensionality and the numbers of holes or handles the particular space contains. As an example, think of the sorts of figures that can be drawn in our own three-dimensional space. A sphere, which is topologically equivalent to a cube, beaker, etc. has no holes and has an Euler characteristic of 2. A torus, with one hole or handle, has an Euler characteristic of zero, as does a cup.

A double torus, or two-handled vase, has an Euler characteristic of −2. In other words, beach balls, cubes, books, and beakers all share a common Euler characteristic and can be deformed into each other. Doughnuts, car tires, and cups have an Euler number of zero, while scissors, teapots, and two-handled vases have an Euler characteristic of −2.

Now this idea of an Euler characteristic, related to the number of "handles" a space contains, is particularly

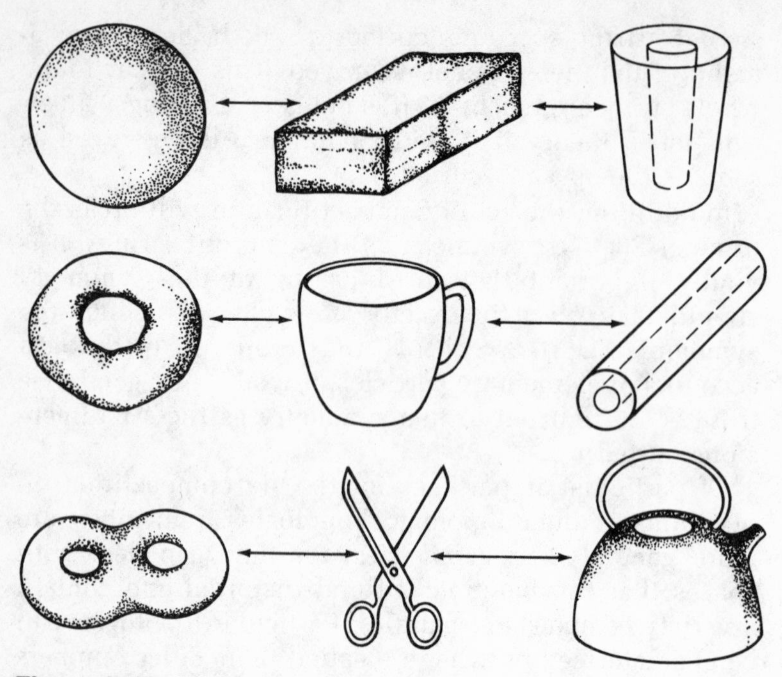

Figure 6–13

By stretching and deforming, it is possible to turn a beach ball into a cube or a beaker, but not into a doughnut. Sphere, ball, cube, and beaker are all topologically equivalent. Figures having a single hole are all topologically equivalent to the doughnut-shaped torus. Another class contains two holes, and so on.

important when it comes to the topologies of compactified spaces. Physicists already know that quarks and leptons occur in a number of "generations"—that is, the elementary particles appear to replicate themselves so that they are in every way identical except for their masses. The reason for this replication was a considerable mystery until the whole business of compactification was understood. Now theoreticians realize that this replication is an inevitable consequence of the topology of the curled up or compactified space. In particular, it has been shown that the number of "generations" of quarks and leptons must be equal to one-half the Euler number of the Calabi-Yau or orbifold space.

The elementary particles observed up to now appear to occur in three generations; thus, the neutrino occurs in three apparently identical forms, called the electron neutrino, the muon neutrino, and the tau neutrino. It has always been something of a mystery as to why nature should replicate itself in this way. The answer, it seems, is that this replication is a natural consequence of the topological properties of the compactified six-dimensional space.

In fact, the Euler number of the Calabi-Yau or orbifold space could be much higher; one reasonable assumption about the compactified space gives it an Euler number of 72, which means thirty-six generations of elementary particles! Clearly this is an absurd result. What is needed is a compactified space with a low Euler number, and this can only come about if the corresponding space contains many holes or handles. But, as we shall see, these holes, in turn, have an important effect on the quantum fields themselves.

Think of a closed string that lives on the surface of a sphere. It is clear that all the loops on this sphere are equivalent; any one of them can be transformed into any other by sliding the loop across the surface of the sphere. There is a high degree of symmetry related to this ability to transform one loop into another.

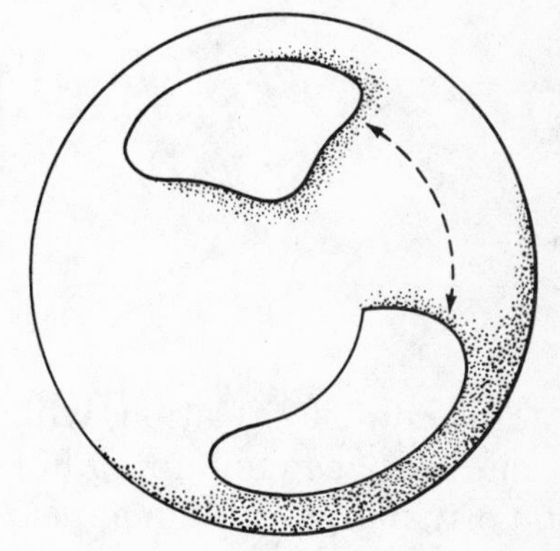

Figure 6–14
It is possible to slide a closed loop anywhere on the surface of a sphere.

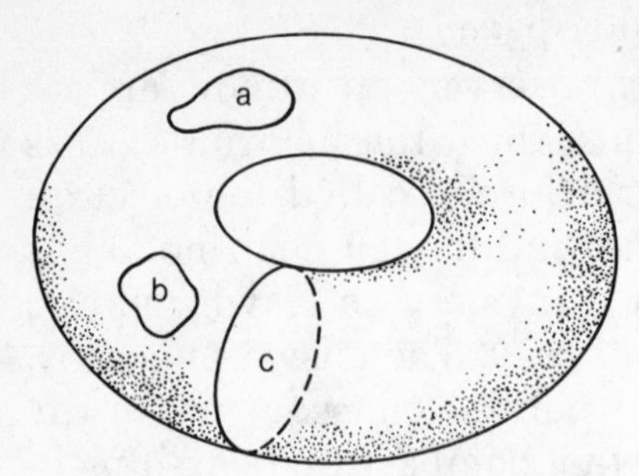

Figure 6–15
By a process of sliding loops, *a* and *b* can be made to coincide, but this can never be done with loop *c*.

Now think of what happens on a space in which there is a hole. It is possible for a string to get trapped by wrapping around this hole. While *a* can transform into *b* by moving the loop around, it can never be transformed into *c*. In other words, there are some loops that cannot be transformed into others. Therefore the presence of holes in an orbifold space has the effect of breaking the basic symmetry of the space. The E8 × E8 symmetry inherent in the original flat ten-dimensional space is therefore broken by the process of compactification. (As we shall see, orbifold spaces also contain what are called "singular points," and it is possible for closed loops to become trapped by these points.)

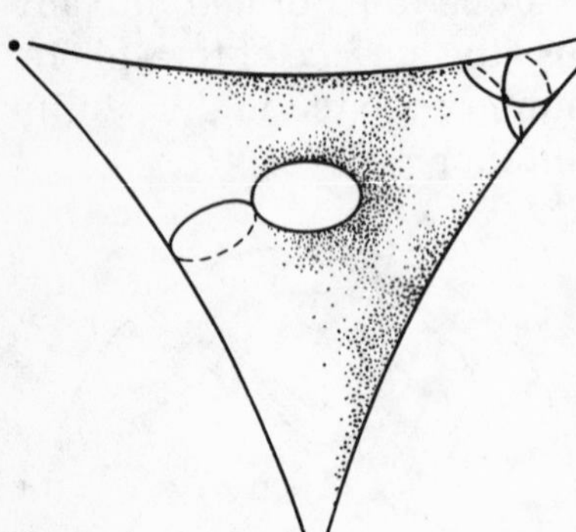

Figure 6–16
In the case of an orbifold, it is possible for a closed loop to become trapped in two ways—around a hole and around a singularity.

This compactification, which involves a Calabi-Yau or orbifold space containing holes, leads to the breaking of the original group, the appearance of the forces and particles of nature, and the requirement that fermions occur in a number of generations.

It is possible that the strings that are trapped by the peculiar topology of the compactified space can manifest themselves in curious ways and give rise to massive particles with fractional charges, or to magnetic monopoles. (A magnetic monopole is a hypothetical elementary particle that carries a single magnetic charge—an isolated north or south pole as it were.)

Therefore, although the six compactified dimensions of superstring space are invisible to us, their effects pervade our world at every level. They are responsible for the appearance and the number of generations of the elementary particles and for the existence of the forces of nature. They may also give rise to a variety of bizarre particles that could one day show up in experiments with very high-energy elementary particles. In fact, physicists are now saying that many of the outstanding problems in modern physics can be reinterpreted as questions about the exact nature of these compactified dimensions.

Conclusion

Gross's theory provides an alternative approach to that of Green and Schwarz and one that many physicists prefer. Yet both theories leave a number of open questions. To begin with, they don't really explain how space compactifies and why we end up with a large-scale universe of four dimensions instead of two, or six, or some other number. And why should this compactification remain stable in a world of quantum fluctuations and in a universe in which the other spatial dimensions are expanding? On both relativistic and quantum theoretical grounds by themselves, we would expect the tightly curled dimensions to change in size. Nor do we have a good explanation for the exact way in which the group E8 breaks into the final unified symmetries. And there is also the question of how the elementary particles achieve their final masses.

There is an incredible jump in energy between the highest energy experiments that physicists can carry out with their particle accelerators and those energies at which the superstring processes are supposed to take place. In fact, this energy gap is as large as that between the elementary particles and our own scale of things! It is not unreasonable to expect to see a whole range of new phenomena and processes before the superstring world is reached. Can it really be true that superstrings are the final solution?

And what of space-time itself? A proper superstring theory should generate its own space-time, since space and superstrings are irreducibly linked. But the best that physicists have been able to do is to put the strings in a flat, inert background space. Moreover, superstring theory, in its present formulation, begins with the quantum theory in its conventional formulation. Should not a theory of everything be expected to change the quantum theory in some equally fundamental way? At such incredibly short distances, will space-time and quantum theory both become transformed?

The present theory leaves unanswered these and a number of other questions. In the final chapter we shall look at some speculative answers, as well as pursue the latest developments and extensions to superstring theory. But first we shall move in a different direction and explore an alternative physics in which extended objects also play the key role. We shall look at twistors, the invention of the mathematician Roger Penrose.

At present the twistor picture of space-time, matter, and interaction is very different from that of superstrings. But some superstring theorists, including Edward Witten, believe that a deeper connection exists—indeed, that the true starting point for a superstring theory of everything should begin with the twistors themselves.

7
From Spinors to Twistors

TO LEARN ABOUT twistors, it is necessary to enter a very different world from that of superstrings. Although at the deepest level there may be a subtle connection between the two approaches, on the surface they reflect very different philosophies. Superstring theory, as we have seen, was built by many hands and did not evolve in any linear or straightforward way. In fact, while the origin of string theory lay in the speculations of the elementary particle physicists, superstrings themselves ended up talking about gravity and space-time in addition to the elementary particles.

Twistors, by contrast, are essentially the work of one man, Roger Penrose, and have their origins not in the mainstream of particle physics but in the more sedate fields of relativity and the mathematics of complex spaces. Yet, like superstrings, twistors also end up having something to say about gravity, space-time, and the possible nature of elementary particles.

While superstrings evolved, like some new animal species, through a series of ideas, modifications, dead ends, false starts, and sudden leaps forward, twistors have had a relatively more straightforward history. The basic approach can be traced to Roger Penrose's interest in the structure of space-time and its meaning in a universe that must also include quantum theory. Assisted by a small and slowly growing band of students and colleagues, Penrose's approach finally flowered to produce powerful insights, not only into the nature of quantum space-time, but also in a number of areas of mathematics and theoretical physics.

Twistors begin with a series of questions that preoccupied Penrose when he was a research student. In particular, he wondered what it means for an electron to spin in an up or down direction. (Remember how, in Chapter 4, we learned that an electron can spin in one of two possible directions.) But what, Penrose asked, would this mean if there was nothing in the universe but one electron? What meaning would it have to talk about two alternative directions in an otherwise empty space? Can spatial directions have meaning in the absence of matter, or do they somehow arise through the actual relationships between material bodies? Penrose was eventually to answer this question by constructing spin networks, a precursor of the twistors.

As a mathematician, Roger Penrose was also interested in the rich and elegant forms of mathematics that are based on complex numbers. But complex numbers themselves are not simply the product of abstract mathematical operations, for they also play a significant role in quantum theory. Possibly complex numbers, he speculated, should enter into the description of space-time in some equally fundamental way, for nature must exhibit a unity of description. Complex numbers and their mathematics were to become an essential feature of twistor theory.

Penrose's training had also taken him into the abstract worlds of geometry, particularly into a somewhat unfashionable field known as projective geometry. Projective geometry could be thought of as a mathematical extension of the relationships that underlie Renaissance perspective and the ways in which one space can be projected onto another. This form of geometry was also to make its mark on the development of twistors.

In fact, powerful geometric visualizations have always been the trademark of Penrose's work. I can remember attending his weekly seminars in 1971, where he would navigate his way through complex spaces in a variety of dimensions, sketching slices and projections of their inner structures and visualizing the lines that twisted and curved through these abstract worlds. And when, more recently, I visited him to check material for these chapters, I found his imagination soaring even higher.

Penrose was born into one of these exceptional English families in which everyone is gifted. His father, Lionel, specialized in the study of genetics and made significant contributions to the study of Down's syndrome. His uncle, Sir Roland Penrose, was a friend of Picasso and a surrealist painter himself. One of his brothers was ten times British chess champion, the other has made distinguished contributions to the field of statistical mechanics, and a sister is involved in medical research on genetics.

As a young man, Roger once went to an exhibit of M. C. Escher's drawings and etchings. Struck by the ways in which the artist had chosen to transform space, Penrose sketched out an *impossible figure,* a three-dimensional object that could have no existence in our real world. The Penrose triangle, as it was later called, was developed by his father into an "impossible staircase," which later became the basis of such famous works as Escher's *Ascending and Descending.* The deeper significance of this triangle will crop up again in these chapters, for it provides a particularly graphic illustration of that branch of mathematics called *cohomology,* which was to form an essential aspect of the twistor approach.

A few decades later, Penrose's interest in spatial symmetry led him to propose a symmetry that had until then not been considered in nature. Physicists and mathematicians knew that crystals can have a twofold, threefold, fourfold, or a sixfold symmetry, but never a fivefold symmetry. While it is quite possible for an artist to draw a

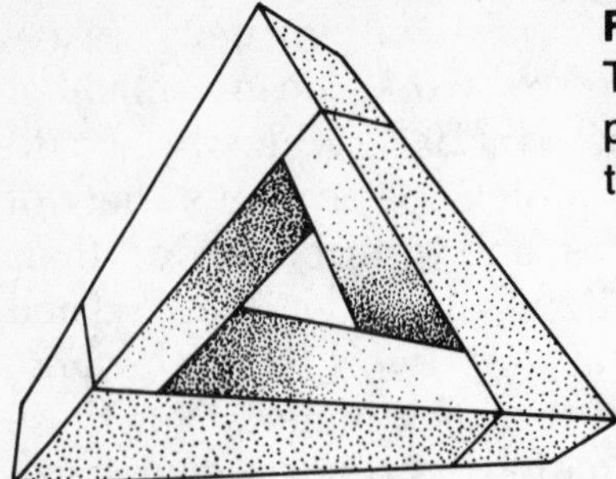

Figure 7–1
The Penrose triangle, an impossible object to construct in three dimensions.

figure with a fivefold symmetry, such a shape, it appeared, could never become the basis for a crystal. The reason is that crystals grow by having the same figure repeated over and over again. But it appeared impossible to generate a lattice out of fivefold objects, because as the individual parts with fivefold symmetry begin to repeat themselves, the various components do not quite fit together, and such a crystal can never grow.

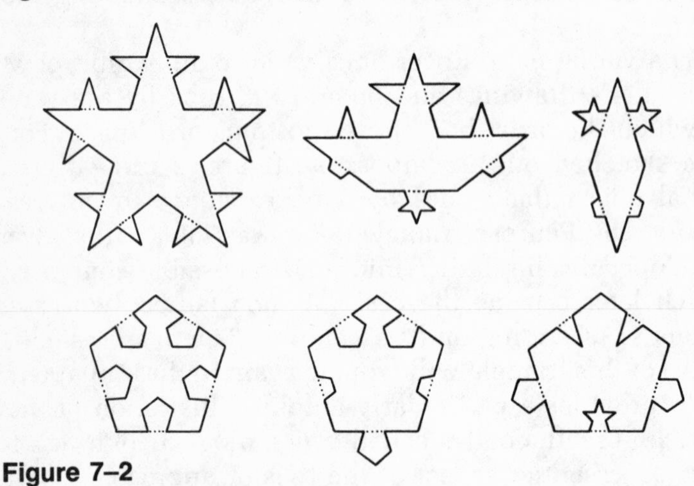

Figure 7–2

Then, in 1974, Penrose came up with a curious way of filling a plane, say this page, with repeating figures—pentagons, rhombuses, pentacles, and "jester's caps." The shapes can cover the page without ever overlapping, a process of growth that can go on indefinitely.

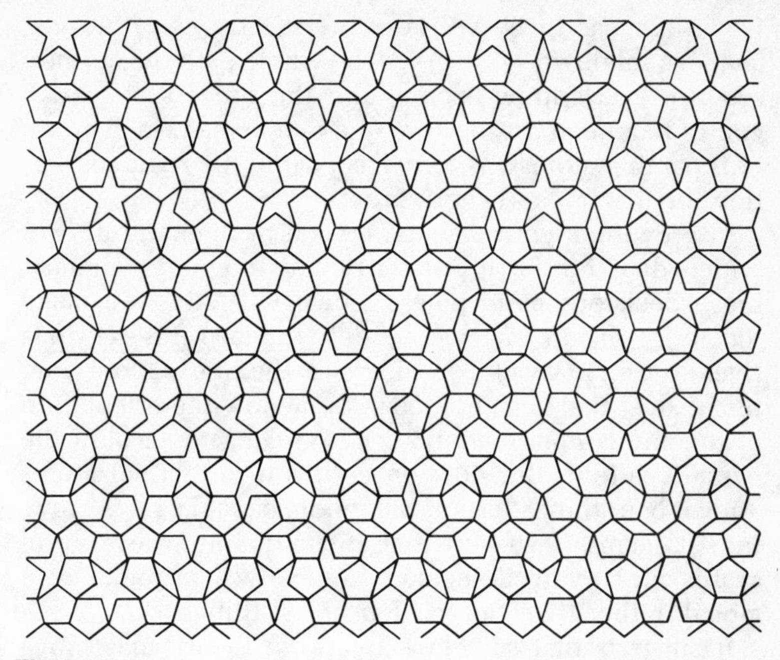

Figure 7–3

Examination of the pattern in the preceding diagram shows that it is based upon a fivefold symmetry that can grow without limit. It is the quasi-fivefold lattice symmetry that physicists had never before taken into account—for fivefold symmetries were believed to be strictly forbidden as the basis for crystal growth. In fact, in 1982, after Penrose had published his ideas, physicists discovered that a crystal of aluminum-manganese alloy had this same pseudo-fivefold symmetry. A quasicrystal had been created. (The symmetry is called *quasi* because the patterns do not exactly repeat; the global structure of the whole pattern is more complex than one based on, say, a fourfold or sixfold symmetry.)

But what has all this talk of quasicrystals and quasisymmetries to do with creating quantum space-time and unifying superstrings and twistors? In a deep sense, there is indeed a connection. The quasisymmetry created by Penrose is subtle; in laying down these figures on the page, one has to be very careful exactly how the next shape is set down—otherwise the whole pattern gets messed up, overlapping occurs, and the process of growth comes to an end. In the case of a real crystal, each group of incoming atoms has to make exactly the right connection, or the crystal cannot grow.

In growing a conventional crystal of diamond or salt, which does not have this particular fivefold symmetry, these decisions as to where atoms will stick are made locally—"on site" as it were. But in the case of the curious fivefold symmetry of a quasicrystal, the rule is more complicated, it has to be nonlocal. Somehow the orientation of incoming atomic groups has to take into account the overall pattern. But how on earth do individual atoms know about the pattern of the whole crystal as it is growing? It seems to suggest that at the quantum level, processes are involved that are somehow controlled at a global, rather than the local, scale of things.

It suggests that somehow quantum theory may move beyond the normal confines of space-time in which objects have specific locations and can only interact with each other via forces. Penrose's quasicrystals point toward a notion of space-time in which distant points can somehow be directly connected, so within each local region there is encoded or enfolded some aspect of a more general global order.

This idea of nonlocality turns out to be essential to the whole twistor picture of things. Moreover nonlocality also seems to be demanded by quantum theory. It appears that our everyday notions of space-time are inadequate, and something much deeper is demanded. We shall return again and again to these ideas of nonlocality as we develop the twistor picture, as well as in the final chapter, which explores some speculative ideas on space-time and the quantum theory.

This preoccupation with the nature of space is in fact the key to all of Penrose's work and explains why the creator of twistors thinks about quasi crystals. It is his belief that the space-time we live in is determined at the quantum level by structures and processes that are non-local in nature. This underlying global structure to our space-time is made explicit in the twistor approach.

Relativity and the Quantum Theory

Roger Penrose's days as a student of mathematics, in the early 1950s, happened to occupy a period during which there was a marked sociological division between relativists and elementary particle physicists. Although Penrose did not become seriously interested in relativity theory until 1958, it is worth recalling the nature of this essential division within the sciences. Elementary particle physics was the mainstream, and most university physics departments had groups of theoretical and experimental physicists engaged in investigating some aspect of the elementary particles. By contrast, relativity was a quieter and more relaxed pursuit in which there was little competition; there were few serious relativists and only a handful of major centers in which large groups worked. Relativity was essentially a theoretical study, and many of its practitioners were trained in mathematics rather than physics.

In those early days, relativity and particle physics had little to say to each other. Of course, the relativists knew that there were serious problems involving the unification of general relativity and quantum theory. The elementary particle physicists, for their part, had little motivation for struggling with the ideas and mathematics of general relativity. Admittedly there were some physicists, like Bruno Zumino at CERN in Switzerland, who realized that the phenomenon of gravity, an essential part of relativity, could not be ignored in the context of a unified model of the elementary particles. I remember in 1970 listening to a lecture given by Zumino on this topic, but he speculated that such unified approaches would have to wait until the twenty-first century. There were also theoreticians who attempted to keep a foot in each field, for they were concerned with quantizing relativity and producing a quantum account of the phenomenon of gravity. But the problems involved in this approach were formidable.

One Hand Clapping . . . One Electron Spinning

In the face of this general lack of connection between relativity and quantum theory, Penrose's early research was concerned with ways in which these two worlds could be brought together. General relativity is a theory about the geometry of space-time, but, Penrose asked, what does this geometry look like in a world that also takes into account the processes of quantum theory?

Some physicists had speculated upon the possible nature of geometry at the quantum level. They had pointed out that the energy within a small enough region of space-time will induce tremendous curvature. John Wheeler had even speculated that, at small enough distances, this space-time breaks apart into a foamlike structure. Other physicists had tried to hold onto a coherent geometry by proposing a shortest possible distance in nature, or that space-time has an underlying lattice structure.

But if quantum fluctuations have a profound effect on space-time structure, then do things also work in the opposite direction? Does the curvature of space-time change even the quantum theory itself? So far, most formulations of quantum theory have been written in a flat space-time. How would the dynamics of a curving space affect the quantum world?

Penrose realized that something radical was called for if quantum theory and space-time were to be united. In fact, he believed that this had to be done, not in our usual space-time, but in a complex space in which the fundamental objects are not points but twisting lines.

Our familiar space-time may not in fact be the background in which the elementary particles play out their lives; rather quantum systems *may define their own space-times!* Penrose's speculation was a particularly bold one, for it suggested that somehow quantum particles are not born into a background space-time, but rather this space-time is created out of quantum processes themselves at the subatomic level. Space-time should not therefore exist before quantum theory but must somehow emerge out of a deeper level.

In fact, this idea had its seeds in the speculations of the eighteenth-century mathematician and philosopher Wilhelm Leibniz, who objected to the assumption of absolute space and time that lay at the base of Newton's new theory of motion. For Leibniz, space arises in the relationships between material bodies. Penrose was now taking this idea a step further by suggesting that quantum systems define their own geometries. If this were true, then it would mean that quantum geometry is important not just at the incredibly small distances of Wheeler's space-time foam, but whenever quantum effects are manifest. It therefore becomes even more important to solve this problem of quantum geometry.

The significance of Penrose's proposals on a quantum geometry can be illustrated by the famous double slit experiment. Photons are sent toward a barrier that contains two slits. After passing through these slits, they fall on the screen behind, where they are recorded. First let's look at the classical case, in which a wave of light is used in place of quantum particles. In this prequantum account of the experiment, physicists would say that part of the light wave goes through one slit and part through the other. Behind the slits, these two wave fronts meet up and create what is called interference, a pattern of light and dark lines on the screen.

Interference is a perfectly general phenomenon where waves of all kinds are concerned. It occurs, for example, when waves meet up after passing through gaps in a

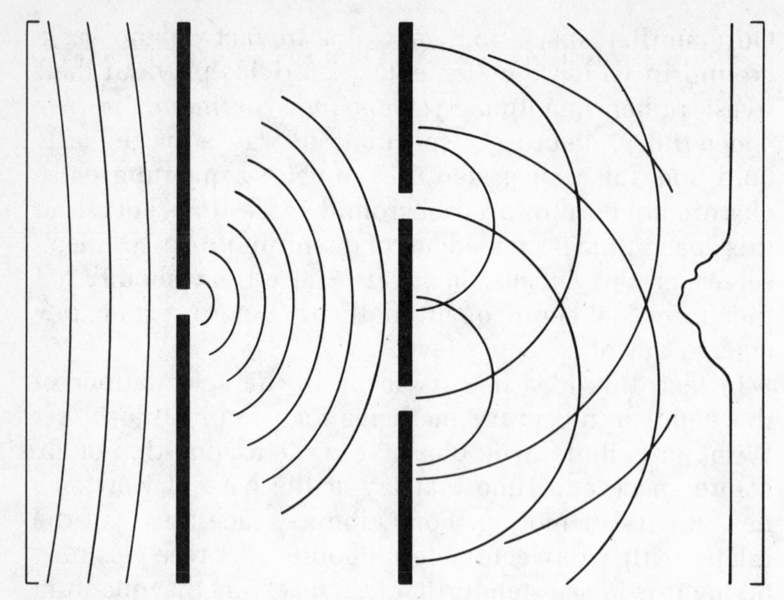

Figure 7–4
After passing through a double slit in a barrier, wavelets spread out and interfere. A complex pattern of light and dark is produced on the screen.

dock. Wherever the waves have to travel different distances in order to meet, in those places they are out of phase. This means that peaks will meet troughs and tend to cancel out; however, a short distance further on, troughs will meet troughs, and peaks meet peaks and reinforce each other. The result is a complex pattern of disturbance.

Such interference patterns occur when light passes through two slits in a screen. However, if one of these slits is blocked off, then light can pass only through the other slit, and since no interference is possible, it simply illuminates the corresponding part of the screen.

In the quantum case, something unexpected occurs, for it is technically possible to cut down the beam of light to the point where only one photon goes through the slit at a time. Common sense tells us that a single photon can only go through one slit or the other. In other words, interference just does not make sense when a

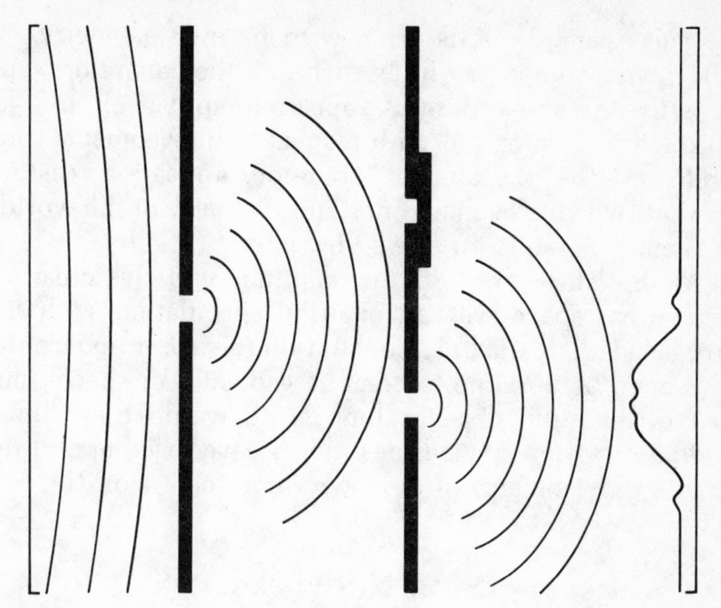

Figure 7–5
Interference cannot take place when one of the holes is covered.

single photon is observed. After all, how could a single photon go through two slits at once and then interfere with itself? Photons are indivisible. Nevertheless the experimental results are quite different. Interference fringes are produced even when the photons pass through the slits one at a time. (In an actual experiment the results of a large number of these single-event processes are recorded on a photographic plate.) But how is this possible? How can a single particle go through two slits and be in two places at the same time? Since this effect was first observed, physicists have attempted a number of different explanations for this quantum paradox.

Penrose's solution to the double-slit experiment is particularly radical. What, he asked, happens if the single photon defines its own geometry? What if, with respect to the photon's space-time, the apparatus appears to have such a curious geometry that it seems to have only a single slit? The photon will pass through this single slit and fall onto the screen. But the screen's geometry will also be so distorted that a complex pattern is built up by successive photons, each with its own geometry. In our own geometry, which includes the laboratory apparatus, the single photon appears to split itself and go through both slits, but with respect to the geometry generated by the photon, the laboratory appears to distort so that two slits become one. The geometry of the world, it seems, depends on one's viewpoint.

At this stage, the idea of quantum particles creating their own space was an exciting speculation with no formal mathematical demonstration to back it up. Would it indeed be possible to create a geometry of space-time out of quantum objects alone? And would these individual quantum space-times then weave together and recreate the geometry of our own large-scale world?

Spin Networks

Penrose was spurred on to create such a new geometry using what he called spin networks. Such an approach, he hoped, would help to answer another question: What is the role of the continuum in physics? This problem had bothered Penrose for some time. Mathematicians claim that there are as many points in this line

---

(the infinite number is named aleph null) as there are in the entire universe. And no matter how finely we subdivide the line into segments, there still remain an infinite number of points between the ends of each segment. This idea of continuity, which occurs both in number and in space, made Penrose uneasy. He was also worried about the way in which continuity crops up in quantum theory when different solutions called wave functions are superimposed.

Why should it, Penrose asked, be possible to divide space without limit into infinitely small parts? After all quantum theory has set a limit on such division when it comes to energy. Light, for example, is not continuous but must ultimately be represented in terms of discrete quantum particles called photons; matter similarly reaches its limits in the elementary particles. Why should space be the exception and capable of infinite division?

Penrose also felt that the universe should somehow be created out of integers alone, using combinatorial processes—that is, simple arithmetic operations such as ratio, addition, subtraction, and permutation. If God is a mathematician, then It creates the universe by counting. In this way space, while appearing continuous at our scale of things, would not be divisible or continuous without limit but would have its origin in finite processes of counting. Some decades later Penrose was to soften this position and admit the significance of the complex mathematics discussed in the next chapter. Possibly the power of complex numbers will eventually be found to provide a complementary picture to counting and combinatoric processes.

But to return to the line. From our perspective it is continuous. Would it be possible, however, to create another form of space though simple arithmetic processes like counting in which infinities and continuities do not crop up? What looks like continuity to us could be the result of the particular energy range in which we live and carry out our elementary particle experiments. If space-time were to have a grainy nature, it could be that our experiments have not yet detected this limit, so that everything still looks continuous. But at much smaller distances, space-time could have a totally different structure.

But what would be the starting point, the fundamental building block of such a space? Penrose decided to start with what is called a *spinor*. The spinor is a mathematical object that is used in the quantum theory to describe the spin properties of the elementary particles. In fact, it is the simplest mathematical object in the quantum theory. By a remarkable coincidence, mathematical spinors also play a significant role in relativity theory. Penrose's intuition led him to the spinor as the possible building block for a space-time.

In fact, he had been puzzling about the meaning of electron spin for some time. His question goes right back to one asked by Ernst Mach, the great nineteenth-century physicist and philosopher whose writings had deeply influenced the young Albert Einstein. Mach not only made significant contributions to the foundations of physics, but his writings also deeply influenced a school of philosophy known as logical positivism; he was the type of philosopher who was always asking troublesome questions.

Mach had been particularly puzzled by the meaning of certain properties, like linear and angular momentum, in an otherwise empty space. In fact, Einstein specifically referred to this question of Mach in the first pages of his great paper "The Foundations of the General Theory of Relativity." What, Mach had asked, would it mean to say that a planet is rotating in space if there were nothing else in the universe against which this rotation can be measured? For example, if the sky were totally empty, how would we ever know that the earth spins on its axis?

It is possible to answer that the earth bulges at the equator because of the effect of centrifugal force, and since we can observe this bulge, it implies that the earth must indeed be spinning. But what exactly is this centrifugal force, and how does it arise? If we choose a set of axes that rotate at exactly the same speed as the planet, then everything would appear stationary in this otherwise empty space. How can a centrifugal force arise in such a case? Where does this force have its origin? What does it mean for there to be a direction for spin when nothing else is present?

Penrose asked this same question in a new way. Quantum theory gives a meaning to the spin of the electron. It claims that an electron can spin in one of two alternative directions. But what meaning would these alternatives have when the universe is totally empty? What difference is there between a spin up and a spin down? We should expect such differences to manifest themselves only when a number of other reference points are present.

So if the distinction between having a spin up and a spin down is to have meaning within a quantum theory set in empty space, it seems to imply that, rather than living in some sort of general background space, electrons actually create their own spaces—a sort of quantum version of our own space-time. Each electron would therefore have associated with it a sort of primitive space—possibly at this stage nothing like our own space-time at all. However, as large numbers of electrons come together, Penrose conjectured, it is possible that their individual protospaces will give rise to a collective space, a sort of shared spatial relationship which may then begin to look something like our own space-time.

Spin, to Penrose, seemed an excellent starting point, and the mathematical building block for spin is a spinor. As was pointed out earlier, this spinor is the most primitive element in quantum theory. Moreover, the quantum rules for putting spinors together involve pure addition and subtraction and have nothing to do with the ideas of continuity. Spinors were tailor-made for Penrose's combinatorial approach.

It is difficult to give a picture of a quantum mechanical spinor, because the spin of an electron is something very different from our classical, Newtonian notion of spin. A ball or a planet can spin on its axis and have a whole continuous range of spins—depending on how fast or slow it rotates. Another way of thinking about the spin of this ball would be to refer to its angular momentum. In the quantum case, however, something very curious happens, for the spin of an electron can have only one of two possible values—up or down. A lone spinor is therefore the simplest possible quantum object, and it can have one of two values. It is a binary object, a single *bit* of quantum space.

But just as a coherent message is built up in a computer

out of many bits, can space be built out of many spinors? The next step, therefore, was to add a second electron, which would join quantum mechanically with the first. At the moment, no idea or property of space is present in the theory; the spinors simply join according to a basic combinatorial rule of quantum theory. Now add a third, fourth, fifth, and so on, building up a whole network of spinors. If Penrose was right, then when it became large enough, this spinor network should begin to develop some of the properties we normally associate with space.

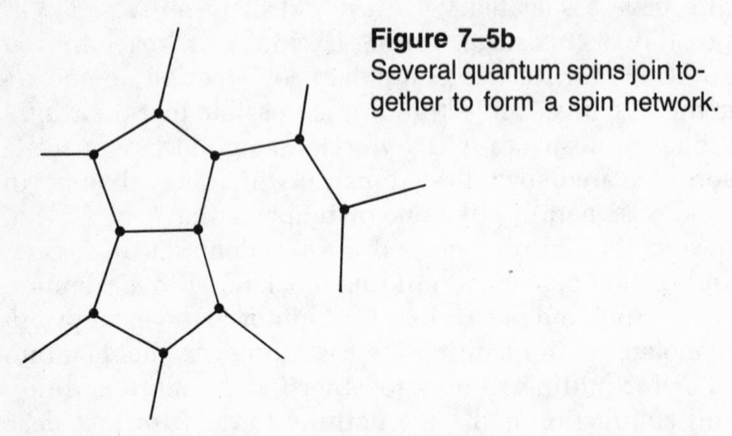

Figure 7–5b
Several quantum spins join together to form a spin network.

Penrose tested his idea by, theoretically, bringing an additional spinor up to this giant spin network. Will there now, he asked, be a meaning to the direction of its spin? To his delight, the spin network answered: direction has meaning, direction emerges out of the relationships between spinors—a relationship, moreover, that is based purely on addition and subtraction. Moreover, the network gave its answer in the language of three-dimensional angles, the same spatial properties one meets in a school textbook of geometry. The spin network had generated the properties of angular directions in a three-dimensional space!

This is a really exciting result, for spinors are two-valued objects yet have created the properties of a space that is three-dimensional. Moreover, this space is generated at the quantum level out of very simple rules of addition and subtraction. Part of Penrose's dream of a combinatorial origin of space had been achieved—this spin network approach involves only the operations of arithmetic and avoids the ideas of infinity and continuity that had bothered Penrose.

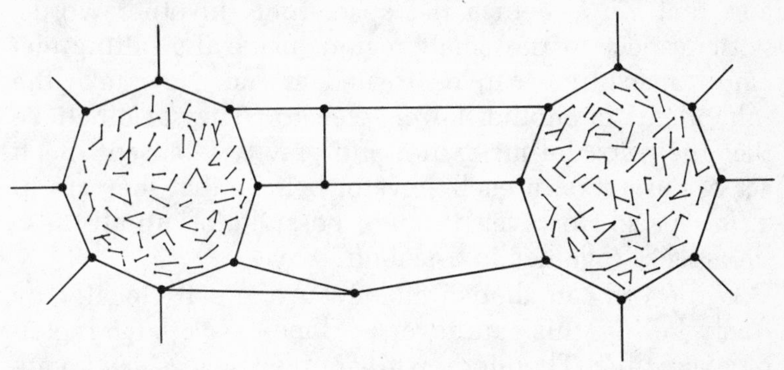

Figure 7–5c
Two giant spin networks meet and interact. The mathematical relationship between them turns out to look just like a Euclidean angle in three-dimensional space.

The next question is what happens when two very large networks are brought together. It turns out that, in general, their respective spaces do not cohere. The result is like a patchwork quilt in which adjacent parts do not match. Is this a failure of the theory? No, it is a marvelous result, for it is exactly how space should look from the perspective of general relativity. General relativity, remember, deals with a space-time that is curved and distorted by the presence of matter.

It is possible to think of a curved space as one that is covered by a patchwork of small, flat spaces. When the overall space is curved, these individual flat segments will not join smoothly at their edges. The effect is similar to trying to cover a beach ball with a series of flat cards—there will always be discontinuities at their edges. But this is exactly what the spin network seems to be suggesting.

This idea can be thought of in another way. What we experience as gravity, the pulling of our bodies to the floor, is the result of living in a curved space-time. But suppose we lived in a freely falling elevator. We would no longer experience any pull to the floor, and all our observations would convince us that gravity is absent and that we live in a flat space-time. In other words, with respect to the small region inside the falling elevator, space-time can be treated as flat. Now take the inhabitants of another falling elevator. They also believe that their space-time is flat and gravity is absent. Each set of observers in each elevator believe that they are in a flat space. However, it is not possible to join all these flat spaces together in a smooth way.

While you can abolish the effects of gravity locally, by freely falling, this can never be done over a large region of space-time. Therefore, while curved spaces can be approximated by a patchwork of small flat regions, these regions will always have discontinuities where we try to join them at their edges. The mathematics involved in fitting regions of space together to cover some larger curved space is called cohomology. The idea of cohomology will crop up again and again in Penrose's work.

In a similar way, Penrose's giant spin networks do not join smoothly. This could be taken to mean that the overall space is curved, or to put it another way, the very fact of this failure to join *is* the curvature of space.

Penrose's spin networks were a provocative concept, but in the end they did not go far enough. They could not in fact be used as a base for the unification of quantum theory with geometry. To begin with, the space they create is incomplete, for it contains no sense of distance or of separation, only Euclidean angles. Moreover, spin-networks create space alone, and not space-time. These spinors are static and nonrelativistic. As a starting point, they are simply not rich enough to create a full space-time. What Penrose needed was a new building block, something that had both a quantum and a relativistic

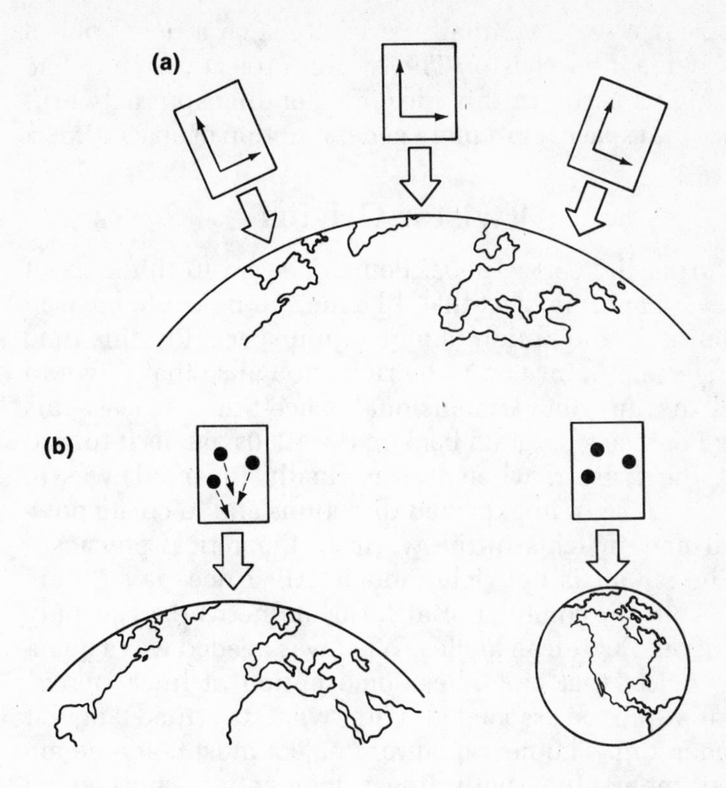

Figure 7–6

(a) The effects of gravity are abolished in a free-falling elevator. Space-time within each elevator appears to be perfectly flat, but clearly these individual flat space-times cannot all be joined together in a seamless way.

(b) Space-time appears flat within a small free-falling elevator. Balls released do not fall but hover in space, maintaining the same positions relative to each other. When the size of the elevator increases, the effects of space-time curvature can no longer be neglected. Since the gravitational field of a planet is everywhere directed toward its center, the balls will experience a slight "tidal force," which causes them to drift together as the elevator falls toward the center of the planet. While space-time many appear flat in small regions, this flatness cannot be extended globally.

nature. He was eventually to create such a novel object and name it the *twistor.* The twistor project therefore had its origins partly in this idea of extending spin networks (quantum spaces) to a more general notion of space-time.

Twistor Origins

During the early 1960s, Penrose began to think about a new quantum object that, like the spinor, could be used to build up the notion of a quantum space. But this time the starting point had to be rich enough so that it would lead to a full four-dimensional space-time. At least this was Penrose's program back in the 1960s, but as it turned out, the twistor, when it was finally invented, was to take Penrose in unexpected directions and to create powerful new insights in the world of theoretical physics.

The spinor is not rich enough. All it does is sit there and spin. In combinatorial form, its networks can only point out Euclidean angles. What was needed was a quantum object that embodies some notion of linear movement. Spin is associated with what is called angular momentum. The new quantum object must combine angular momentum with linear momentum, and on an equal footing. It must be an object that is both spinning and moving along. In addition, it must be both quantum mechanical and relativistic. Penrose's twistor was to fulfill all these requirements. It was also to bring together a number of other key ideas that Penrose had been working on: the significance of complex numbers and their geometry; the role played by light rays in relativity (also called null lines), and the special ways in which physical solutions are singled out in quantum field theory.

All these ideas were to meet in twistors, and from these twistors would flow new intuitions on the nature of quantum fields, the internal structures of the elementary particles, a theory of quantum spaces, and new mathematical results in several important branches of mathematics. By the end of the 1960s, twistors had been created, and for the next two decades, their implications were unfolded in a variety of directions. Today one of the major researchers into the topic of superstrings, Edward Witten, is suggesting that these same twistors may be the true starting point for superstrings.

To understand the full significance of what Penrose had created, it is necessary to know something about the power of complex algebra and why complex geometry may be the most natural way to make the connection between quantum theory and the general theory of relativity. To enter the world of twistors, it is first necessary to open the door of complex numbers.

Complex Worlds

The algebra of complex numbers, and the geometry of the spaces they generate, is a powerful and elegant branch of modern mathematics. Not only is it a subject of study for mathematicians, but it also plays a significant role in physics, particularly in quantum theory. It was for these reasons that Penrose became convinced that twistor theory must be based within that field which mathematicians call complex analysis. Today Witten and other superstring scientists are also working in complex analysis, the mathematics that unfolds from the properties of complex numbers.

Working with complex numbers enables theoreticians to call upon another important tool, cohomology. A combination of the very powerful mathematics that stems from complex numbers and cohomology has enabled Penrose to create new maps for the world where quantum theory and relativity meet.

The essential idea of cohomology can be understood by returning to the Penrose triangle, that "impossible figure" that was to inspire some of Escher's etchings. Suppose that you were to observe the triangle, section by section, through a telescope. You would see each of three segments that, taken together, make up the entire figure.

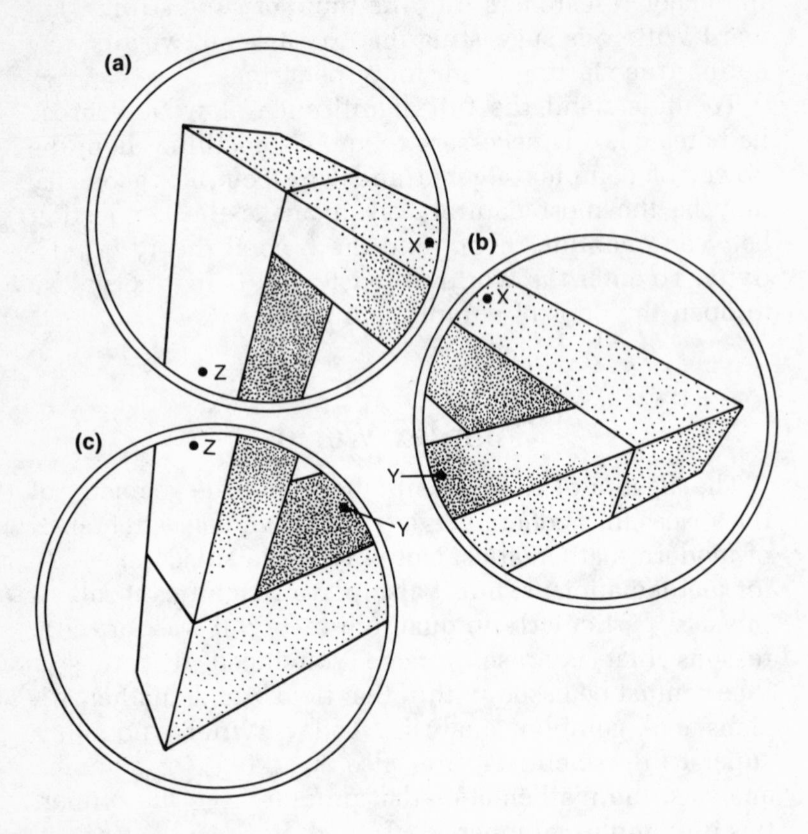

Figure 7–7
The corners of a Penrose triangle (Figure 7–1) are observed through a telescope and appear to make physical sense. An observer attempts to combine these three views and to make sense of them. But how can the common point *X*, in the views (a) and (b), be made to coincide in such a way that the common points *Y* and *Z* also coincide? While portions of the Penrose triangle make sense locally, their global properties defy spatial logic.

Segment (a) looks perfectly reasonable. There is no ambiguity in the figure, and it would be perfectly possible to construct it out of two pieces of wood. The same applies to Segments (b) and (c). Each of the three segments is a perfectly proper figure. Since these figures overlap slightly, it is possible to identify a point x that is on both the segment (a) and the segment (b). Likewise the points y and z can also be identified as common to other pairs of segments. Finally, we try to build the complete Penrose triangle by moving the segments toward or away from us until they all join up. This means that x, y, and z on the segments must match exactly. Of course, in the case of an impossible figure like the Penrose triangle, this can never be done. While each segment makes sense by itself, when taken together they cannot join smoothly to form a sensible figure.

Another way of seeing this is to realize that the point x must be a different distance from the viewer depending on whether it is defined in the segment (a) or (b). Yet x is supposed to be a single point common to both segments. Clearly the figure must be paradoxical.

The properties of the Penrose triangle illustrate the general approach of cohomology. This topic deals with the properties of spaces that can be divided into a series of small regions, each of which is well behaved. In general, however, it may not be possible to join all these regions together smoothly and unambiguously. Cohomology therefore enables physicists to work with complicated, distorted spaces by dividing them into overlapping regions, each of which is, on its own, well behaved.

We shall return to cohomology again and again during the next two chapters. But first it is necessary to move on to the complex numbers, for Penrose's twistor theory and its cohomology is defined in what is known as a complex space.

What exactly are complex numbers? You may remember that they have something to do with the square root of -1, that mysterious number that is also called i, the imaginary number. In fact, perhaps without realizing it, you have already been exposed to the power of complex numbers and complex analysis. The dual resonance model was created to explain how elementary particles scatter off each other during high-energy experiments. Since the effects of the strong nuclear force were too strong to solve directly, physicists resorted to a fundamental property of the S-matrix, which is itself a complex function. Without having to solve a number of very difficult equations, they were able to learn something about the S-matrix by relating what are called its *real* and *imaginary* parts. Significant deductions could then be made about the elementary particles based on the mathematical good behavior of complex functions.

One way of thinking about the complex numbers is that they allow algebra to continue in the face of a certain class of equations. Look at the following three equations:

$$x^2 + x - 6 = 0$$

$$x^2 - 9 = 0$$

$$x^2 + 1 = 0$$

At first sight they look more or less the same, but there is something very strange about the third one.

The first, $x^2 + x - 6 = 0$, can also be written as $(x + 3)(x - 2) = 0$. For the whole expression to vanish, either $x + 3 = 0$ or $x - 2 = 0$. In other words, the solution to this equation is either $x = -3$ or $x = 2$.

Likewise, $x^2 - 9 = 0$ has two solutions. It can be written as $(x + 3)(x - 3) = 0$, indicating the solutions $x = 3$ and $x = -3$.

The third equation, $x^2 + 1 = 0$, runs into problems when we try to write it in the form $(x + ?)(x - ?) = 0$. This can only work if ? is replaced by $\sqrt{-1}$. Yet there seems to be no way to define the square root of a negative number; no such number exists in the early history of mathematics.

But what if we assume that $\sqrt{-1}$ is a perfectly legitimate number in its own right? In fact, why not define a whole continuum of imaginary numbers? If these imaginary numbers are allowed into mathematics, then it becomes possible for algebra to go on to solving equations like $x^2 + 4 = 0$, $x^2 - 2x + 4 = 0$, etc.

In addition to the real number line, in which numbers are placed on a line as follows:

$$\ldots \quad -3 \quad -2 \quad -1 \quad 0 \quad 1 \quad 2 \quad 3 \quad \ldots$$

Why not have a number line for the imaginary numbers?

In such a line, imaginary numbers add and subtract the way real numbers do: $2i + 3i = 5i$, $6i - 3i = 3i$, and so on. Addition and subtraction keep us on the number line. But what about multiplication? $2i \times 3i = -6$, since $i \times i = -1$. Multiplication takes us from the imaginary number line to the real number line.

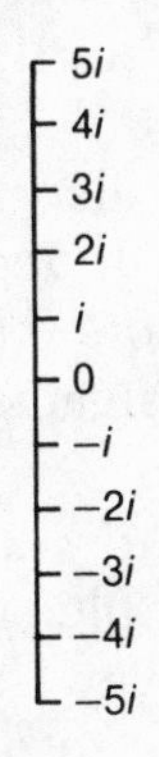

Figure 7–8
The imaginary number line

The obvious next step therefore is to make the whole thing consistent and combine the two lines. In place of two number lines, there will be a number space. Numbers that actually lie on the axes are either pure imaginary or pure real numbers. But off these axes, each number is a mixture of real and imaginary parts. A number such as $3 + 2i$ is called a complex number and is therefore represented by a point in the complex number space.

One advantage of using complex numbers is that they enable us to locate any point in a plane using a single

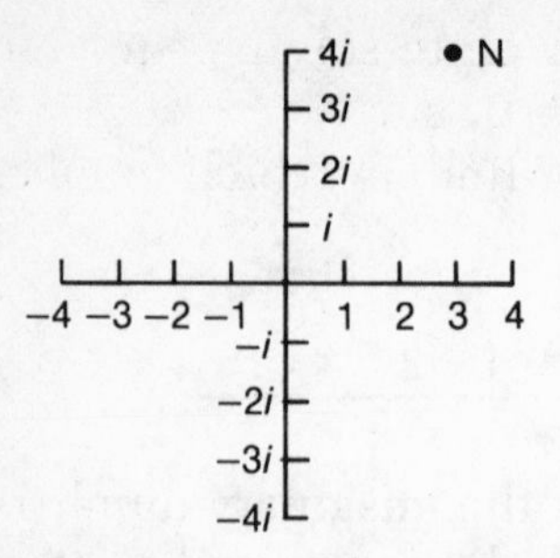

Figure 7–9
By combining the real and imaginary number lines, it is possible to create a space of complex numbers. The point N corresponds to the complex number $(3 + 4i)$.

complex number and therefore provide an alternative scheme to the more familiar Cartesian coordinates. You may have seen photographs of a modern mathematical application of this idea, called the Mandelbrot set. This is related to the general topic of fractals, and, in this case, rich and infinitely complicated images can be generated in a simple way starting from a single complex number.

Complex numbers are also used when physicists and mathematicians work with vectors. With their help, the rules for combining vectors become straightforward; indeed, there is a whole algebra of vectors based on the complex numbers.

But defining points and vectors in terms of complex numbers is relatively simple. The real power of complex numbers comes in the way they can generalize algebra and geometry. Being able to describe a system in terms of functions of a complex rather than a real variable has tremendous advantages. To begin with, there are formal relationships between the real and the imaginary parts of such functions which stem from their mathematical good behavior (called "complex analyticity"). Just as a complex number always has a real and an imaginary part, a complex function can be similarly divided. For example, the relationship between the real and imaginary parts of a function determines how the absorption of light, as it passes through a piece of glass, is related to the way this light is bent and broken down into its component colors.

What is truly amazing about this relationship is that it follows directly from the mathematical behavior of complex functions and not from the physical details of a particular piece of glass. There are a host of other important physical relationships between the real and imaginary parts of a complex function. The dispersion relations of the S-matrix are a case in point. They tell us about the relationship between elementary particle resonances and scattering experiments simply by relating the real to the imaginary parts of the scattering matrix. In many instances, mathematicians and physicists find that by pushing a function into complex regions—"analytic continuity" as they would put it—they can bring more powerful techniques to bear. Clearly if Penrose's God is a mathematician, then while It may have created the universe by counting, It certainly had the beauty of complex geometry in mind when It did so.

But if this complexity is so important, why do we never see it directly? The length of something is 2 feet, never $2i$ feet or $2 + 3i$ feet. Likewise, the gasoline in your car is measured in gallons, not complex gallons. It appears that while complexity underlies the physical world, each time we try to see it by making a measurement, this complexity hides its face.

Mathematicians have a formal recipe for getting real results from complex numbers. Every complex number N has its mirror image reflected in the real axis. This is called its complex conjugate N*.

Figure 7–10
The complex conjugate of N $(2 + 3i)$ is N^* $(2 - 3i)$; it is obtained by reflecting N in the real axis.

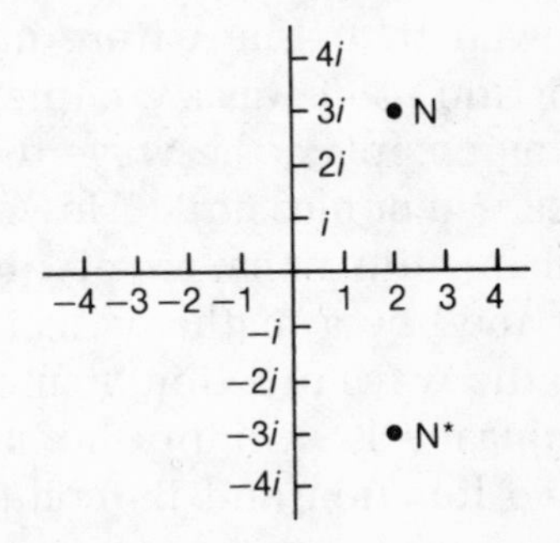

The complex conjugate of the number $2 + 3i$ is $2 - 3i$. Therefore, whenever a number is multiplied by its complex conjugate, the result will be real, for $N \times N^* = (2 + 3i)(2 - 3i) = 4 - 9i^2$. And, since $i^2 = -1$, the result is $4 + 9 = 13$, a real result obtained from two complex numbers.

Whenever a complex number and its conjugate pair up, the result is always real. Or to put it another way, it could be that underlying our real universe there are products of complex numbers and their conjugates. This, as we shall see, is exactly what happens in quantum theory. So mathematical complexity is always hidden from us by the pairing up of complex things with their conjugates to give real answers.

That complexity must underlie our real world can be most powerfully seen in the case of the quantum theory. In 1925, a matter of weeks after Werner Heisenberg had discovered quantum mechanics, Erwin Schrödinger came along with an alternative approach. At first sight, Schrödinger's equation had a direct intuitive appeal, for it looked like the sort of equation that describes waves in water or air. This differential equation of Schrödinger was called the wave equation, and its solutions were called wave functions. All observed quantities can be calculated using the wave equation and turn out to be real.

At first it seemed as if electrons in an atom could be thought of as real standing waves—like the vibrations of a string. But closer examination of the mathematics showed that things were not that simple, for the wave function itself was a complex rather than a real function. Being complex, the wave function can never be directly observed or measured. However, the physical and observable quantities that are predicted by the theory are always obtained by a mathematical operation that involves taking the wave function $\bar{\Psi}$ and multiplying by its complex conjugate $\bar{\Psi}^*$. Simply taking the product of a complex wave function and its complex conjugate, for example, gives the probability of locating a quantum particle in a given region of space. But the product of a complex function and its complex conjugate must always give a real result!

So although the underlying formalism of the quantum theory is complex, its predictions always involve real numbers. Again complexity hides itself under the cloak of reality. Penrose felt that such complexity is so fundamental to the quantum world that it must enter explicitly into the description of quantum space-time. In generalizing his spin network to twistors, it would therefore be necessary to work in complex spaces, and the whole power of complex mathematics would have to be used.

Null Lines

Another key idea in Penrose's approach is the light ray, also called a *null line*. In fact, this null line is also connected with the idea of complex numbers. You will recall from Chapter 5 that, because the transformations of Einstein's theory act to mix together space and time coordinates, Hermann Minkowski decided to introduce time on an equal footing as the fourth dimension in physics.

But this fourth coordinate does not quite enter as the other three. It is not written as t but as $-ict$, where c is the speed of light. That is, it enters as an imaginary number. But we never see the imaginary or complex side of space-time, because all measurements turn out to be real. Take, for example, the distance traveled on a journey through space-time. The distance on any graph is calculated by first measuring the distance along each of the axes. For example, to find the distance AB on the following graph, we measure three units on the x axis and four on the y axis. The total distance is then obtained using the Pythagorean theorem.

But what happens when one of these axes is the time axis? Suppose we measure the distance taken by a rocket

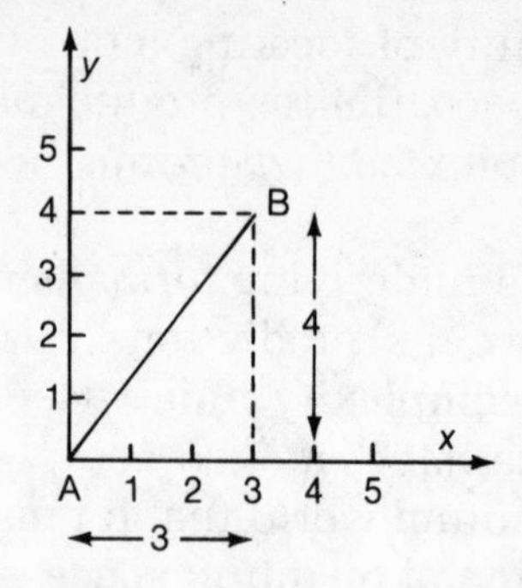

Figure 7–11
The distance AB can be calculated using the Pythagorean relationship, $AB^2 = 3^2 + 4^2$. The result is $AB = 5$.

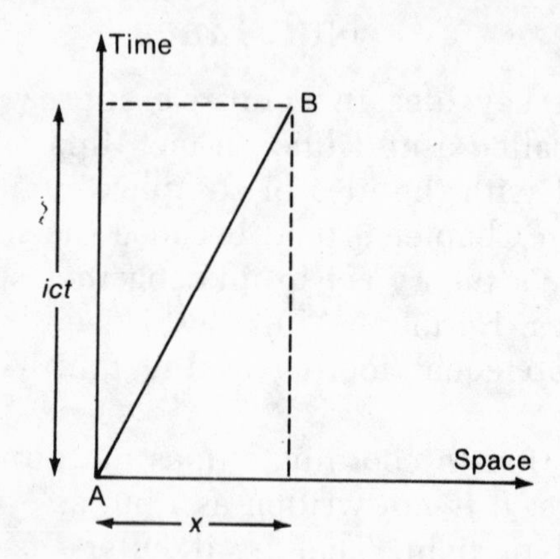

Figure 7–12
The Pythagorean relationship is also used to calculate distances in space-time. In this case the time coordinate appears as *ict*. Hence:

$$AB^2 = x^2 + (ict)^2$$

$$AB^2 = x^2 - c^2t^2$$

When $x = ct$, the measure of the distance AB becomes zero. This condition specifically holds for anything that moves with the velocity c.

on its journey through space-time. This distance is $\sqrt{i^2 - c^2t^2}$. Provided the rocket moves at speeds less than that of light, this distance has a positive value. The result is a real number, and the imaginary nature of the fourth coordinate does not show up.

But what happens when our rocket moves at the speed of light? In this case, $x = ct$, and the distance is *zero*. We have what appears to be a finite line, yet it has a length of zero. This is one of the famous null lines of relativity. (Null lines are to take a key role in twistor theory.) They are the tracks taken by massless particles, which move at the speed of light. (Note that the speed of an automobile or rocket is given by distance/time. This means that when the rocket moves at a speed v, the spatial distance x that it will travel in t seconds is vt. It is possible to substitute this value of vt for x in the space-time distance equation. The equation therefore becomes:

$$\sqrt{v^2t^2 - c^2t^2} = \sqrt{t^2(v^2 - c^2)} = \text{distance}$$

which also indicates that when $v = c$, the space-time length vanishes.)

But what does it mean to say that the length of a null line is zero? Traveling at constant speed on a highway, we can measure our journey in hours—it is five hours from the city to our vacation destination, two hours to our summer cottage. Now, in the theory of relativity, we must always refer times and distances to particular observers. Take, for example, what happens to the internal time experienced by an elementary particle that is traveling very close to the speed of light. When high-energy cosmic rays strike the earth's upper atmosphere, they collide with the nuclei of oxygen and nitrogen atoms and create a whole cascade of "secondary particles"—mesons and the like whose speeds are very close to the speed of light. But some of these mesons are very short-lived when measured in the laboratory. Knowing how long these particles have to live before they disintegrate, and how fast they are going, it is a simple matter to calculate that they can travel only a short distance before decaying. Nevertheless, these particles are registered in laboratories on the earth's surface after traveling for several miles through the atmosphere. How is this possible? The answer is that while, with respect to a stationary observer, the particles live only for a very short time, according to their own internal clocks, they live much longer and are therefore able to travel for many miles in their lifetime. Another way of looking at this is that, with respect to the speeding particle, the distance from the earth's upper atmosphere to its surface is only a few feet, and not many miles.

What happens if we go even faster—if a clock could be carried on board a null line? (Of course, a material body like a clock cannot move at the speed of light, but our argument is hypothetical.) In fact, no time would have elapsed between leaving *A* and arriving at *B*! All distances would have shrunk to zero. For light, not even one second ticks away from the time it leaves a distant galaxy until it reaches your eye. Look up at the night sky and realize that, along a light ray, the distance to the stars is zero. When you look along a null line, nothing separates you from all that you see in the universe around you!

Of course, we generally say that a star is so many light-years away. This means that, for us as stationary observers on the earth, the light takes several years to travel from the star to earth, or that the information that now reaches us from the star concerns events that happened on the star's surface several years ago. However, with respect to light itself, this time interval is zero, and the distance vanishes. This is a direct consequence of the fact that time, in relativity theory, enters as an imaginary quantity. In fact, the structure of a beam of light in space-time, what relativists like Roger Penrose would call the *light cone structure,* is most naturally expressed in terms of complex spinors.

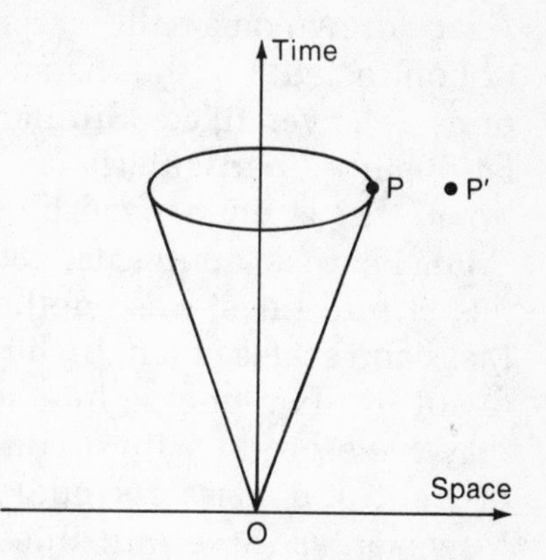

Figure 7–13
The light cone, a key geometrical feature in relativity theory. Light from *O* spreads out in space with increasing time. Note that the point *P* can be reached from *O*, but not the point *P′*. *P′* cannot therefore be causally connected in any way to *O*. Events at *O* and *P′* can have no influence on each other.

Null lines and the speed of light occupy a very special place both in relativity theory and in twistor theory. That lines can have the special measure of zero length is a direct consequence of the fact that time enters space-time with an imaginary rather than a real factor. Complex numbers enter physics at many levels in relativity. For example, they appear in certain of the solutions to Einstein's field equations. Penrose therefore felt that they must form the foundations of twistor theory.

Null Lines and Conformal Geometry

Null lines, or the paths taken by light and massless particles, are of such importance in relativity that Penrose has suggested *only null directions are really "there."* Indeed, it is possible to create a geometry based on null lines alone. But in a universe having such a geometry, mass would not exist, for only massless particles can move at the speed of light. Since null lines always have zero length, scale and distance would have no meaning in such a universe. Change the scale of this universe, and all physical quantities remain the same. The property of being unchanged under changes of scale is called *conformal invariance.*

Conformal geometry is the sort of geometry that exists on a balloon that is being inflated and let down again. A face drawn on a balloon remains a face as it is stretched or contracted, yet the distances between the various features is never fixed. Similarly there is a rich set of relationships, particularly as regards the causal connections, that is unchanged by conformal transformations.

For Penrose, conformal geometry, based on the properties of null lines, was another key to quantum geometry. Mass and scales of length, he conjectured, are not primary quantities but emerge in a secondary way. The universe may have begun with a conformal geometry, a universe of light and massless bodies moving along null lines. However, as these null lines began to interact, the basic conformal invariance was broken and mass was born into the universe.

A Space That Divides

By the early 1960s, Penrose was thinking seriously about a complex geometry that would be built of null lines and would employ all the power of complex mathematics. This geometry would also connect with the basic ideas of spin networks. With the help of such a geometry, Penrose hoped to be able to make deep interconnections to quantum theory. The question was, how to proceed?

The answer lay in his creation of twistors, objects that lie partway between relativity and quantum theory and, being more general than spinors, combine linear and angular momentum and live in a world of complex dimensions. Indeed, this space of twistors may well be the primordial space of the first elementary particles. In the early years of twistor theory, Penrose had also hoped that this twistor space could be used to generalize his original notion of a spinor network and produce a full four-dimensional space-time. But this project ran into serious difficulties. Nevertheless, the twistor space itself was to provide powerful new insights.

Penrose was also concerned about a fundamental problem in quantum field theory. It concerned the way in which physicists have to pick out real, physical solutions in quantum field theory. Quantum field theory is a generalization of the quantum theory of Heisenberg and Schrödinger and deals with the quantization of nature's fields. Light, for example, is a property of the electromagnetic field, and in quantum field theory this electromagnetic field must be quantized. The electromagnetic field's various quantized vibrations are discrete quanta that can be identified with individual photons of light. In a similar way, one can write down for matter a quantum field whose excitations are elementary particles such as electrons.

The problem with this mathematical formulation is that it allows for a whole spectrum of solutions having both positive and negative frequencies. But only half these solutions, the ones with positive frequencies, show up in nature. An elegant theory should not leave its last step unfinished. If quantum field theory is to explain the origin of quantum particles, then it must also tell how to pick out the physical, positive frequencies. Penrose, for his part, felt that this principle of selection should be built in a geometrical fashion. In the field of complex numbers, for example, the real axis bisects the complex space and divides it into a positive and a negative part. Would it be possible to create a space of quantum field solutions that is similarly divided, in a very natural way, into a positive and a negative-frequency part?

By the fall of 1963, these various ideas were running through Penrose's head: an extension of the spinor that formed the basis of his spin network, a conformal geometry based on null lines, the power of complex functions and complex geometry, and a space that would have a fundamental division into a positive and a negative part. At that time he was staying with the famous relativity group at Austin. Following a weekend break in San Antonio, the group was driving home when Penrose began to play around in his imagination with a diagram he had recently constructed.

A friend of Penrose, the physicist Ivor Robinson, had been wrestling with a difficult problem involving the electromagnetic field. On his way toward creating a solution, Robinson had recently come up with an algebraic expression but did not know exactly what this result would look like. Penrose had, for his part, worked out a geometrical visualization, a complicated picture using a series of null lines (or light rays). Now, as he relaxed on the drive home from his weekend vacation, this configuration came back again into his mind. He was struck by the idea that such a diagram demands a space of complex dimensions. Moreover, this space would naturally divide itself into two regions. Such a division, Penrose realized, must be related to division of the quantum field into physical and unphysical solutions.

Suddenly all the pieces of the puzzle began to fall into place. Penrose realized that what he had been looking for was a complex space built out of complicated congruences of null lines. It would underlie the space of quantum theory, the primordial space in which the first photons moved, a space in which a single quantum of curvature would have meaning. Just as the matter around us is built out of primitive quantum entities, so too spacetime would be derived from this more elemental space. It would be a space built not out of points but of twisting congruences created out of straight null lines. Penrose would go even further; the lines themselves, these basic twistors, are not only the components of geometry but of the elementary particles as well. Twistors are the generalization of quantum mechanical spinors. They are dynamic and not static objects, for not only do they have angular momentum, but linear momentum as well. These twistors are extended objects, and the space that is built out of them must now be seen in an entirely new way. In twistor space, lines are fundamental; points are only

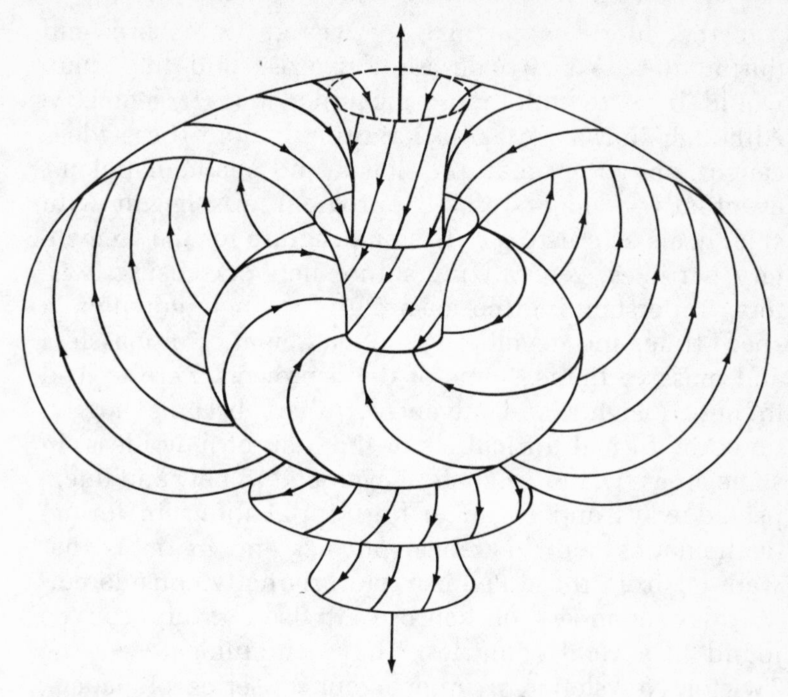

Figure 7–14
A visualization of the Robinson congruence in terms of null lines

secondary objects created through the intersection of lines. In a twistor theory we may expect the whole idea of points to take second place and be replaced by non-local descriptions of space-time.

On the one hand, twistor space would contain the rich conformal geometry of null lines; on the other, it is related to the physical and unphysical solutions of quantum field theory. Suddenly a fundamental property of the quantum field had been directly connected with an equally fundamental geometrical and relativistic property of space-time. The idea was exhilarating. The twistor was a Janus-like object, unified, yet with one face pointing toward quantum theory and the other toward general relativity.

In that fundamental insight, twistors were born, and during the next two decades, Penrose and his group would have to unfold the meaning of their geometry. Although it was to take a long time before these ideas caught on with other physicists and mathematicians, eventually a major research program was begun with such goals as creating a twistor structure for the elementary particles, generalizing spinor networks using twistors, understanding the meaning of a single graviton in space-time, and developing the mathematics of massless and massive fields. Some of these projects were to lead to new insights and advances; others became bogged down by formal difficulties or for want of new ideas. In some areas twistor geometry developed in new and unexpected ways and began to find applications in formal mathematics and theoretical physics and in fields that were far from those Penrose had originally considered.

Today members of Penrose's twistor group can be found in several countries. They communicate via the Twistor Newsletter, a mimeographed series of papers, notes, and abstracts, some of them handwritten. The cover of the newsletter bears a drawing by Penrose of a group of twistors that represent the Robinson congress—the original image that had started the whole thing in the early 1960s.

The story of how twistor geometry and the mathematics of twistor space were unfolded will continue in the next two chapters. In the final chapter, we shall explore some possible connections between twistors and superstrings and learn how both approaches are really taking us into some very deep questions, questions that may one day lead physicists to a supertheory in which space-time structure and quantum theory both take on new forms.

8
Twistor Space

PENROSE'S TWISTORS ARE part of what the physicist John Wheeler has called "the great dream," a vision that all of physics can be reduced to geometry. In the nineteenth century the great dreamer was an English mathematician and philosopher, William Kingdon Clifford. Clifford had been investigating the new geometries of Bernhard Riemann and Nikolay Ivanovich Lobachevsky that went beyond the more familiar schoolbook geometry of Euclid. Speaking to the Cambridge Philosophical Society on February 21, 1870, Clifford said:

> I hold in fact (1) That small portions of space are in fact of a nature analogous to little hills on a surface which is on the average flat; namely, that the ordinary laws of geometry are not valid in them (2) That this property of being curved or distorted is continually being passed on from one portion of space to another after the manner of a wave (3) That this variation of the curvature of space is what really happens in that phenomenon which we call the motion of matter.*

---

*By a stroke of irony, it was Clifford's other major contribution to mathematics, the Clifford parallel, that first led Penrose to his fundamental twistor picture. While puzzling over the meaning of Ivor

For Clifford there was nothing in the physical world but this variation of geometry, these little hills and changes of curvature. Clifford's dream was later taken up by Einstein, who extended the notion of space to space-time and demonstrated that the geometry of space-time determines the motion of matter and is, in turn, governed by the surrounding energy and matter. But Einstein wanted to go further, to complete this dream of Clifford and demonstrate that matter itself is nothing more than the knots and hills of space-time, and that fields of force are also regions of curvature. If the force of gravity can be reduced to geometry, he speculated, then why not the magnetic and electrical forces?

This quest preoccupied Einstein in vain during the last decades of his life as he attempted to modify his basic equations of general relativity in an attempt to make additional room within the geometry of space-time for matter and force. Hermann Weyl, his contemporary, tried a slightly different approach. In Einstein's theory, lengths and the rate of clocks can change, but if an object is taken on a circuitous route and is finally brought back to its initial point, its length should be unchanged. Weyl, however, allowed this length scale to vary from place to place but, at the same time, introduced a sort of standardizing field that enabled this varying scale to be fixed at each point—by a fundamental length gauge as it were. This gauge field, he argued, would look exactly like an electromagnetic field but would inhabit a space-time in which scale can vary.

While Weyl did not ultimately succeed with this idea that the electromagnetic field is a gauge field, the notion did, however, surface in a very different form many decades later, when force fields were treated as gauge fields in elementary particle theories. The 1920s also saw the

---

Robinson's solution, Penrose was led to a geometrical construction in terms of what are known as Clifford parallels. It was while he was thinking about this diagram that Penrose was first struck by the idea of twistors.

attempt of Theodor Kaluza and Oskar Klein to increase the dimensions of space-time from four to five and in this way try to portray the electromagnetic field in terms of the effects that would appear as this extra dimension is curled up.

But, in the last analysis, this dream of Einstein and his contemporaries could never work, for it took no account of the great advances that were being made in quantum theory. Despite his significant contribution to the birth of quantum theory, Einstein felt uneasy with its final form and could not accept the theory as a complete account of nature. Yet quantum theory claims to offer a full description of the ultimate structure of matter. So how can a theory like the one Einstein was searching for, which purports to explain matter in geometric terms, ignore this complementary quantum mechanical description? Some physicists proposed to take account of this by trying to give space-time a quantum structure. They attempted, for example, to add in rules of quantum theory to a curved space-time. But such research projects always ran into serious difficulties.

In addition, the experiments of elementary particle physicists had added to physics two new forces, the weak and strong nuclear force. Einstein's unified theory would have to take account of both of these, in addition to electromagnetism and gravity. But these new forces are essentially quantum mechanical in nature, and the sorts of theories that Einstein and his contemporaries were working on were purely classical.

Clearly a new unified account of nature demanded something much deeper. Roger Penrose, for his part, believed that a true geometrical account of nature would have to involve a unification of quantum theory and space-time. But this could never be done using some simple extension of the classical space-time of Einstein and Minkowski with its four real dimensions and its foundation in the dimensionless point.

For Penrose, the starting point had to involve a space

of *complex* dimensions, since complexity also occurs in a fundamental way in quantum theory and, in addition, complex geometry is far richer than real geometry. A complex space, he believed, would be able to account not only for the space of everyday, macroscopic objects but the space of quantum processes as well, and offer a geometrical account for photons and a single quantum of curvature. It would be the primordial space that eventually gives birth to our own space-time.

The basic objects of this space are not points but one-dimensional objects that are extended in space. More general than points, they are at the same time quantum mechanical in nature and elements of a complex geometry. By the late 1960s, Penrose had found his space, this space of the great dream. It is a space of complex dimensions, and its basic objects are twistors. These twistors can be thought of as generalizations of light rays or null lines and are defined by complex numbers. And, although twistors can be pictured in a space-time of real dimensions, their real home is twistor space.

Just as matter has its foundations in the elementary particles, so space-time will ultimately have its origins in twistor space. Twistor space becomes the new arena of physics, the ground on which quantum processes are played out. Clifford's great dream was of a physics reduced to the geometry of space-time. Now Penrose had transformed that dream: the proper arena of physics is not space-time but the complex dimensions of twistor space.

Complementary Pictures

Is it possible to give a picture of the twistor and an image of twistor space? That rather old-fashioned topic, projective geometry, which Penrose had studied as a student, now comes to our aid, for with its help it becomes possible to go back and forth between two pictures, one in twistor space and the other in space-time. First let us examine the underlying twistor picture.*

Since the twistor, which we shall call Z, is built to have both angular and linear momentum, it is more general than a spinor. It can be defined by complex numbers, which are in fact its coordinates in twistor space. A twistor Z therefore becomes a point in twistor space. Being complex, the twistor Z also has a partner, its complex conjugate called Z^*. As we would expect from complex numbers, the result of multiplying the twistor Z by its complex conjugate Z^* is a real number; in fact such a product can be used to define s, the helicity or degree of twist of the twistor.**

$$\tfrac{1}{2}Z \cdot Z^* = s$$

This helicity turns out to be an important factor. It can be positive, negative, or zero. As we shall see, twistors with zero helicity have a special role to play in space-time, for they look exactly like rays of light. Since twistors are built to have a positive, zero, or negative twist, this means that they also have a natural right- or left-handedness—which is also called chirality. Chirality, therefore, is totally natural within the twistor picture and does not become a major problem for physicists as it did in the early string approach.

All the twistors with zero helicity, $Z \cdot Z^* = 0$, lie in a special region of twistor space which we shall label

*The twistor space we shall be mainly exploring in this book is built with three complex dimensions and is called projective twistor space. There is also a full twistor space having four complex dimensions. But the important properties and arguments of this and the following chapter can be derived from the three complex dimensions of projective twistor space alone. We shall only need to refer to full twistor space again in Chapter 9—when we set up a very curious space-time in which the wave function for a single graviton can be written down. For the moment, however, let us forget about the space of four complex dimensions and concentrate on the projective twistor space of three complex dimensions.

**The symbol · between the two twistors indicates multiplication.

PN. This region *PN* has the effect of dividing twistor space in half.

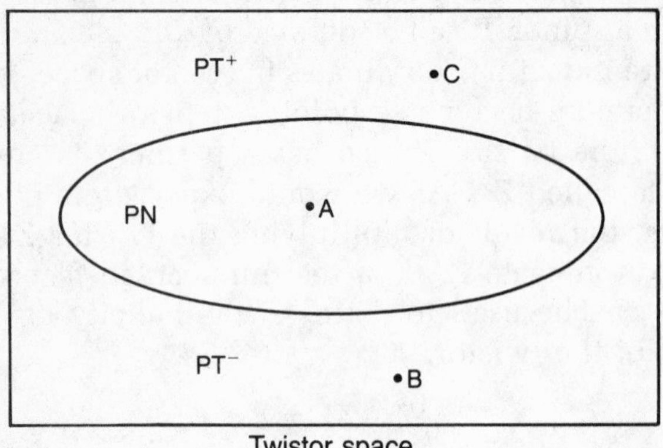

Figure 8–1
PN divides twistor space into two regions, PT^+ and PT^-. This division is the geometrical analogue of the way in which solutions in quantum theory are divided into positive- and negative-frequency parts.

Above *PN* can be found the region PT^+, which contains only twistors of positive twist. Below *PN* lies PT^-, whose points correspond only to twistors with a negative twist or helicity, while all the points on *PN* itself are the coordinates of twistors with zero helicity. This division of twistor space in half by the region *PN* turns out to have a profound physical meaning. Remember how Penrose had been interested in the way physics must pick out the real, or physical, solutions from a quantum field. This process of selection involves picking out positive-frequency solutions from negative-frequency ones, and it had always struck Penrose that this should have a purely geometrical interpretation. But now, working in twistor space, this becomes possible. The selection of solutions in quantum field theory is related to whether points lie in PT^+ or in PT^-. We shall also find, when the twistor picture is extended to include curvature, that the space-time of individual graviton wave functions is also broken down into curvatures with a right- or a left-handed sense.

Having established twistor space, whose points are the twistors and which is divided by *PN* into two parts, it is now possible to create the complementary picture in terms of the more familiar space-time of special relativity, more specifically the flat space-time first created by Minkowski. It turns out that twistor space has a far richer structure than the flat Minkowski space-time of Einstein's theory of special relativity. Not only is it possible to recover or reproduce this space-time out of twistor space, but a whole structure of geometric relationships can be recovered as well. While the traditional approaches have emphasized points and other local features, the essential strength of twistor space will lie in its ability to describe large-scale—also called global—structures in space-time.

Take, for example, a point in that special slice of twistor space that we have labeled *PN*. Points in *PN* represent twistors with zero twist, and it turns out that they correspond to light rays or null lines in space-time. In fact, by taking all the points in this section (*PN*) of twistor space, we can fill space-time with null lines or light rays.

Since the twistor is the most fundamental aspect of the geometry of twistor space, we now see that the corresponding basic object in space-time is not a point but a null line. Space-time is to be recovered from the more fundamental twistor space, and its foundation becomes lines rather than points. Its geometry is based upon twistors, which look very like light rays or the tracks made by massless particles. In fact, there is a fundamental duality between the null line in space-time and a point in projective twistor space *PN*. One way of thinking about this is that the local structure, the points, of twistor space encode global or large-scale information about space-time.

If these extended null lines are the essential geometric objects of space-time, then what about points? These are now secondary or derived objects that are defined by the

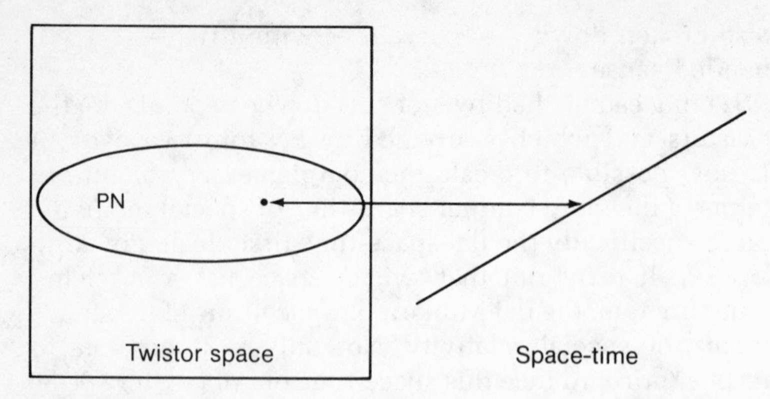

Figure 8–2
A point in the PN region of twistor space corresponds to a null line (a light ray) in space-time.

intersection of twistors. By way of an example, think of an interchange on a highway. This is represented by a point on a map, but it could also be thought of more globally as something held in common by several different routes. In other words, these routes have their meeting at this intersection. From the perspective of a traveler in a car, who is concerned with journeying from state to state, the various routes are of key importance, and the intersection is a secondary thing used for getting from one route to the next. Intersections are therefore perceived from a global perspective.

Likewise, a point in space-time could be thought of as a secondary thing given by the intersection of null twistors or light rays. A point is, in fact, defined by the collection of null twistors that pass through it. But this means that a point is therefore nonlocal in its deeper nature. The origin of space-time now appears very different when viewed from the twistor perspective.

It is also possible to go in the other direction and look at a complementary picture. Points in that special region *PN* of twistor space correspond to twistors of zero helicity—that is, light rays—in space-time. On the other hand, points in space-time will now correspond to extended structures in twistor space. In fact, points in space-time define lines in that special region *PN* of twistor space. The duality is complete: Points in *PN* correspond to null lines in space-time. Points in space-time correspond to lines in *PN*, a special region of twistor space. Yet the twistor picture is more fundamental, for we take the points in twistor space as the start of geometry, while the points in space-time are secondary objects. Space-time appears to be fundamentally nonlocal.

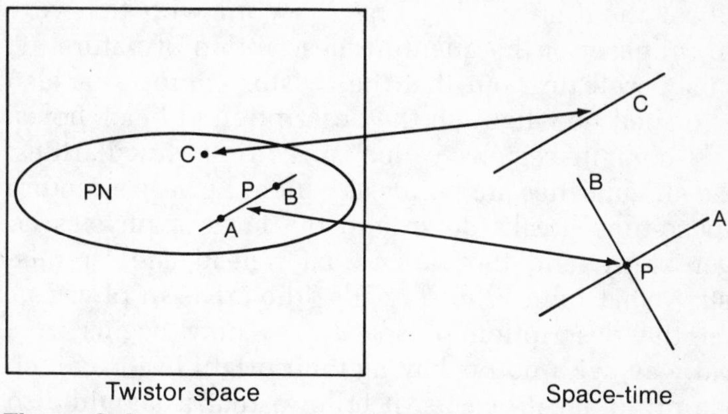

Figure 8–3
While the points *A*, *B*, and *C* in twistor space correspond to lines in space-time, the line *P* corresponds to a point in space—the intersection of the lines *A* and *B*.

Twistors are essentially nonlocal objects concerned with the global structure of space-time; points in space time are of secondary importance. This aspect of the twistor picture has some immediate implications. To begin with, it opens the door on a totally new way of understanding space-time and the elementary particles: relationships are now to be viewed at a global rather than a local level.

Nonlocality, for example, enters into the quantum picture of nature in an essential way. It was Niels Bohr who first stressed that quantum mechanics is essentially a nonlocal theory. Indeed, one way of avoiding paradoxical interpretations about quantum measurements is to assume that quantum theory refers always to nonlocal descriptions. A famous thought experiment suggested by Albert Einstein, Boris Podolsky, and Nathan Rosen can only be properly understood if nonlocal connections are assumed between different parts of the quantum system—that is, distant connections that do not involve the causal action of any force of a physical field. (This whole topic of nonlocality will be treated in greater detail in the author's next book.) Since twistors are fundamentally nonlocal themselves, they are in accord with this very basic property of the quantum description of nature.

It may well turn out that the twistor approach is also able to shed new light on the description of black holes, which contain regions called space-time singularities. These singularities are points at which the very structure of space-time breaks down and the laws of physics no longer apply, and they are a major headache for relativists—what John Wheeler calls "the crisis in physics." Since the description of space-time now begins in a global way, with points having their origin in the coincidence of nonlocal objects, it follows that it should also be possible to have a nonlocal description of a black hole.

The black hole singularity occurs where matter and energy collapse down to a single point in space-time. At this same point, the basic structure of space-time is also supposed to break down. But now it becomes possible to begin with a global description that refers to all space-time including the singularity itself. In this way, it may be possible to retain a description of a black hole in which the laws of nature need not break down. Even more importantly, this description could include the first singularity of the universe—the big bang.

The fact that space-time points are secondary objects, derived from twistor intersections, also implies that we need not expect them to survive when quantum processes are admitted into the twistor picture. In effect, certain transformations or processes in twistor space turn out to be equivalent to quantum processes in space-time. And, as with any transformation, the basic geometrical units become interchanged. In this case, a quantum transform of twistor space mixes up the twistors. But since points in space-time are defined in terms of the conjunctions of these twistors this means that the space-time point will smear out. At the quantum level, the twistor space picture suggests that points in space-time lose their distinction and become "fuzzy."

Let us now continue to develop our complementary twistor and space-time pictures. Twistor space, of course, contains far more than that special region, *PN*, of zero twist (or helicity). (Technically *PN* is called a submanifold of projective twistor space.) What then, is the meaning of the points that lie above and below *PN*, in the regions we will call PT^+ and PT^-?

What sort of geometrical picture do they create in space-time? Since the geometry and structure of twistor space is far richer than that of our four-dimensional space-time, we cannot expect the points in PT^+ or PT^- necessarily to have a simple geometrical picture. To put this in another way, simple structures in twistor space may correspond to rich and complicated structures in space-time. Remember that points in PT^+ or PT^- will denote twistors with positive or negative helicity, and it turns out that these geometrical structures cannot be pictured in terms of single null lines alone.

A twistor with positive helicity has to be represented in the space-time picture by a collection, called a congruence, of null lines that twist around each other in a right-handed sense. Similarly a twistor in PT^- becomes a set of null lines twisting around each other in a left-handed sense. Therefore, while points in *PN* correspond to null lines in space-time, points in PT^+ or PT^- correspond to twisting congruences of null lines. Conversely, null lines, or light rays, in space-time correspond to points only in a particular section, *PN*, of twistor space, while the more complex twisting patterns of lines in space-time correspond to the other points in twistor space.

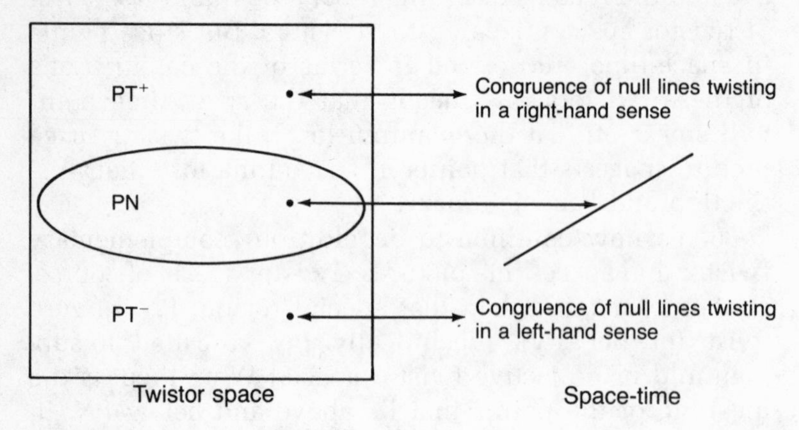

Figure 8–4
A general point in twistor space does not correspond to any simple geometrical picture in space-time.

Twistor space has been used to re-create space-time, along with a structure of twisting lines. In a sense, twistor space is much richer than space-time; it contains far more information, since out of its point we are able to create complicated geometrical patterns of lines in space-time. This is one consequence of the fact that twistor space has complex dimensions—since a complex number is made of two parts, it contains more information and is more flexible than a single real number. Likewise, projective twistor space, with its three complex dimensions, has a much greater geometrical potential than space-time alone. Space-time has always been a backdrop to physics, and structures have always been added to it and lines drawn in it. But with twistor space, this space-time is re-created along with its various structures. It is impossible to separate the null lines, the geometrical representations of light rays, from the space-time in which they move.

The Twistor Program

Having created twistor space, Penrose was now able to map out his research program. It was to occupy his group—first at Birkbeck College, London, and then at the Mathematical Institute of Oxford University—over the next two decades. Goals of this program included:

- Extending the earlier ideas of spin networks and generating a space-time out of twistor relationships alone
- Expressing the elementary particles, their internal structures and symmetries in terms of twistors
- Using the complex analytic properties of twistor space to understand the various fields of physics
- Exploring the implications of quantum theory for the twistor picture and speculating on ways in which quantum theory may be transformed
- Understanding how space-time curvature enters via twistor space and giving a rigorous treatment of quantum gravity

While certain aspects of this program, such as the twistor networks, have proved enormously difficult, others have led to remarkable insights and have had important spin-offs in other areas of mathematics and theoretical physics. We shall explore some of the accomplishments of the twistor program in the rest of this chapter, leaving the whole question of quantum gravity and massless fields to Chapter 9.

Superstrings and Twistors

At this juncture, it is a good idea to make a direct comparison between twistors and superstrings. Superstrings are massless, one-dimensional objects having an incredibly short length. Twistors, as null lines or light rays, have no length, no sense of scale, and no mass.

Superstrings are defined in a ten-dimensional space, which, it is assumed, will compactify down to our own four-dimensional space-time. Twistors, by contrast, are defined in a space of complex dimensions. This complex twistor space is then used to generate our four-dimensional space-time, along with its rich structure of null lines.

Superstrings carry a series of internal symmetries, which are broken as the ten-dimensional space compactifies. Such symmetry breaking does not necessarily occur in the twistor picture, rather—as we shall see in Chapter 9—certain symmetries such as that between right-handed and left-handed photons and gravitons are violated at the outset. At the same time the "twist" of the twistor means that its approach is basically chiral.

As closed loops, superstrings have a natural interaction with the vacuum of space, which can be pictured in terms of the creation and annihilation of gravitons. In the twistor picture, gravity and quantum processes are interconnected in a different way.

In its present formulation, superstring theory accepts the basic formulation of the quantum theory. Penrose, however, believes that this must change in the twistor approach so that quantum theory and space-time have to be described in new ways.

There are also a number of subtle connections between superstrings and twistors. Penrose's colleagues, Lane Hughston, William Shaw, and Mike Singer have shown that a relativistic string corresponds to a general curve in twistor space. This means that relativistic strings can be derived from the twistor space picture. In addition, certain important twistor transformations can be generalized to the ten-dimensional space of superstrings. More recently a connection has been made between the trouser diagrams of string theory and twistor diagrams. Researchers in the superstring and twistor fields are currently investigating the same rich fields of complex geometry and complex analysis, called cohomology theory, which may reveal yet deeper connections between the two approaches. Indeed, Edward Witten believes that the twistor formulation may be the proper starting point for superstring theory.

Mass in a Universe of Light: Conformal Invariance, the Light Cone, and Twistor Space

A point in twistor space corresponds to a complex twisting structure of null lines in space-time. And, in special cases, when the point lies in the region *PN* that bisects twistor space, it corresponds to a single null line or light ray. The complex geometry of twistor space can therefore be used to generate a complementary space-time picture full of twisting congruences of null lines. Indeed, the null line now occupies the key position in geometry. A consequence of this use of null lines is that the corresponding space-time is said to be conformally invariant. That is, the space-time contains no sense of a basic scale. All the relationships discussed so far are totally indifferent to the scale of length. It is as if the space-time built from twistor space is drawn on the surface of a giant balloon that can grow or diminish in size. While distances are in a constant state of flux, other, more important structures are unaffected—for example, circles in such a space always remain circles.

You will recall from the previous chapter that Penrose gave great emphasis to this property of conformal invariance when it came to setting up his twistor space. Now it turns out that conformal invariance, being a property of null lines or light rays, is connected to a very important space-time structure that relativists call the *light cone*. Switch on an electric lamp, and light spreads out to fill the room. This spreading is like the ripples created when you throw a stone into a pond, except that light is a spherical wave that expands in the full three-dimensional space. Now translate this into the space-time picture. In place of the surface of a sphere that expands in time, we have a four-dimensional cone. A cross-section of this cone in one spatial and one time dimension is shown in the following diagram. It represents the way light spreads from a point O.

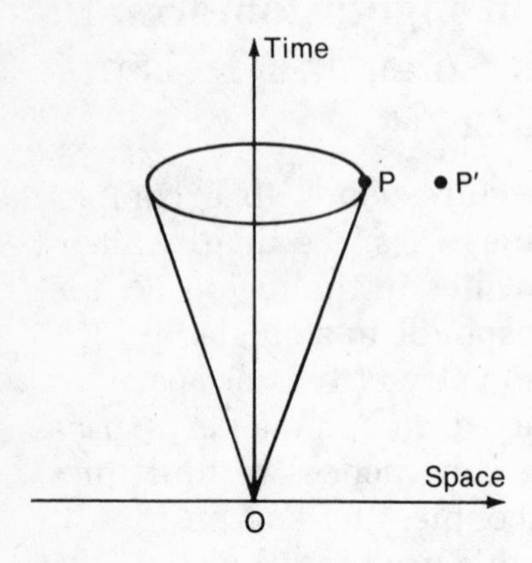

Figure 8–5
The light cone, a key geometrical feature in relativity theory. Light from *O* spreads out in space with increasing time. Note that the point *P* can be reached from *O*, but not the point *P*′. *P*′ cannot therefore be causally connected in any way to *O*. Events at *O* and *P*′ can have no influence on each other.

Note that point *P* (which represents an event occurring at a particular point in space and time) is reached by this beam of light. This means that an observer at the space-time event *P* is able to see the light from *O*. In fact, all points inside the light cone are causally connected. But look at the point *P*′. The only way it could be connected to *O* is via some particle, or transmission of information, which goes faster than light. Points, like *P*′, that lie outside the light cone generated from *O* have no causal connection to *O*. They cannot be affected in any way by an event at *O*.

The light cone is of key importance in relativity, since it determines the whole causal structure of interrelationships.* Conformally invariant spaces always leave the structure of this light cone unchanged, since the changes

*This general structure of the light cone also has connections with quantum theory—light cones are most naturally defined using spinors, the same spinors that also crop up in quantum theory!

of scale associated with conformal transformations leave this light cone unchanged. The light cone structure is therefore said to be invariant (unchanged) by conformal transformations, and any transformations within twistor space will also preserve the light cone in space-time. But Penrose was also concerned with keeping a meaning to the light cone even when quantum fluctuations are allowed into space-time. How this can be done will be explained later in this chapter.

Conformal invariance was the keystone of Penrose's original program. As we have seen, only null lines like light rays or other massless objects are allowed into this picture. But how then were massive objects, such as the elementary particles, to be pictured in space-time and twistor space? Penrose argued that mass is really a secondary quality, something that is built up through the interaction of more fundamental massless, conformally invariant objects. But this means that the appearance of mass has the effect of breaking the basic conformal invariance of the theory. In addition, massive elementary particles cannot correspond directly to a single (conformally invariant) twistor but will have to be represented by collections of null lines.

But there is yet another way of breaking conformal invariance, and this is to introduce certain nonlinear interactions such as gravity. Now gravity, in Einstein's general relativity, is associated with the appearance of general curvature in space-time, and it is just this curvature which breaks the conformal invariance of space-time.

To sum up, there is a powerful interconnection between, on the one hand, conformal invariance and a massless, scaleless space-time built from free, noninteracting twistors and, on the other hand, interaction, mass, scale, general curvature, and gravity interaction. Exactly how these various relationships are to be understood in terms of the geometry of the corresponding twistor space has become a major research topic for Penrose's group. For when conformal symmetry is broken in space-time, this means that the geometry of twistor space must have changed in a significant way. Understanding the full nature of these problems will probably involve a powerful connection between general relativity and a theory of the elementary particles.

Twistors and Curvature

In the mid-1960s, following his initial insights into twistor space, Penrose had some difficulties in persuading his contemporaries of the power of this new approach. Being relativists, they were used to working in the curved space-times associated with gravity, for it is in the curvature of space-time that all the riches of general relativity lie. But the space-time generated from twistor space is (conformally) flat. This seemed, at the time, like a backward step. However, Penrose was trying to generate a picture of space-time curvature at the quantum level itself, one that would emerge through a series of quantized elements of curvature. The long-term goal, of course, was to generate something that looked like the curved space-time of general relativity but that would have purely quantum origins.

Earlier on in this chapter, we showed how twistor space can be used to produce a complementary flat space-time picture. There is a striking duality between points in a special region of this twistor space and lines in space-time, and vice versa. It is therefore natural to ask if this picture can be generalized to a space-time that is curved by the passage of a wave of gravity.*

But at the very first hint of curvatures, all the power of the original twistor picture seems to disappear! As soon as the slightest curvature is introduced into the flat space-time, the corresponding beautiful mathematical structure of twistor space looks as if it is lost. Curvature

---

*Remember that in Einstein's general theory of relativity, gravity is equated with the curvature of space-time.

seems to be incompatible with all that rich and powerful mathematics.

A closer inspection, however, shows that, rather than the twistor picture being destroyed, something more subtle has taken place. In fact, the complex analytic properties of twistor space are transformed in a special way. It turns out that this change can be turned to our advantage, for it suggests a way of understanding the intimate connection between gravitational curvature of space-time and quantum processes.

Introducing curvature in the space-time picture produces transformations of the points in twistor space. In other words, it mixes up twistors and their complex conjugates. But this mixing or interchange does not happen haphazardly, rather it preserves certain basic relationships between the twistors and their complex conjugates. It turns out that each twistor and its conjugate are interchanged in precise ways. While twistors are interchanged, certain vital twistor relationships are nevertheless preserved.

It is now possible to ask what happens when a wave of gravity or curvature passes through a flat space-time. A twistor and its complex conjugate become interchanged, and this looks suspiciously like what happens during a quantum process! In Chapter 5, it was pointed out that quantum theory describes processes, measurements, and change in terms of what it calls *operators*. It now appears that the twistors themselves behave in similar ways to quantum operators and that the passage of a gravitational wave looks a little like an actual quantum process.

Heisenberg's uncertainty principle tells us that the act of making a measurement of position followed by a measurement of momentum gives a quite different result from first measuring momentum and then position. Translated into more formal terms, it tells us that the process of using a momentum operator followed by the position operator gives a different result than if the position operator is applied first and then followed by the momentum operator. In fact, actual quantum processes can look like the ordered application of operators.

But now we see that in twistor space a twistor and its complex conjugate look something like a momentum and a position operator. Since the passage of a wave of curvature through space-time produces an interchange of twistors and their conjugates in twistor space, what happens begins to look very much like quantum processes in twistor space.

In other words, a wave of curvature in space-time looks like a quantum process in twistor space. Conversely a quantum transformation in twistor space looks like a wave of curvature in space-time. The duality between the two pictures is achieved by connecting gravity and curvature in space-time to quantum processes in twistor space.

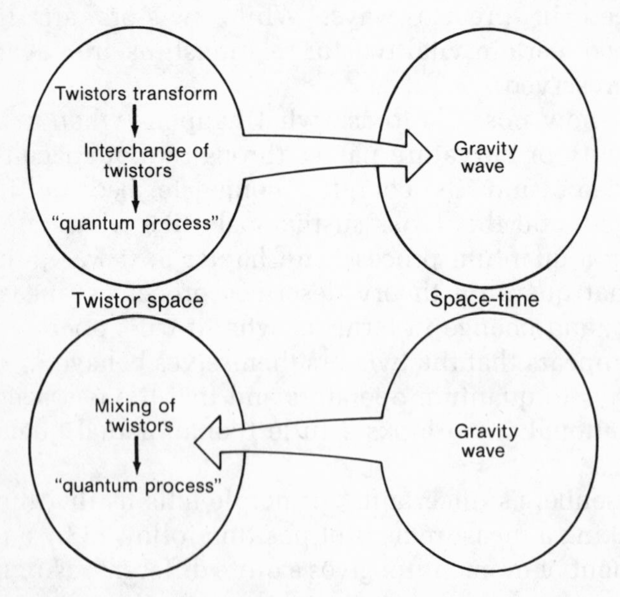

Figure 8–6
An elegant symmetry is established between what look like quantum processes in twistor space and a gravitational or curvature wave in space-time.

This suggests that there may be a deep connection between general relativity, gravity, and space-time on the one hand, and quantum processes on the other. It even suggests that one theory may be affected by the other, so that taking account of gravity may actually change the quantum theory itself. If Penrose's conjectures are right, then the twistor approach may be pushing physics in a much deeper direction.

There is another way of picturing these ideas. Think of a flat region in space-time. All its null lines and geometry are well defined. Now move to another flat region in which null lines are similarly well defined. What happens when we attempt to join these two regions? Once a wave of curvature has been allowed to pass through space-time, it becomes impossible to join these two flat regions exactly.

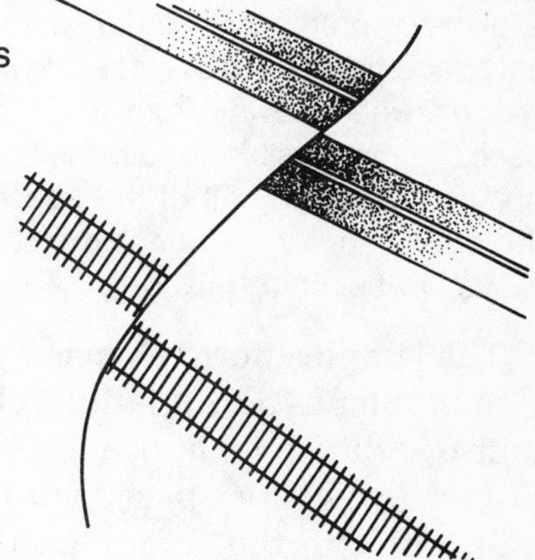

Figure 8–7
The effect of an earthquake is to shear roads and railroad tracks.

The result is a little like what happens when a shock wave from an earthquake passes through a town. Driving down the main street, we suddenly discover that it has been shifted a few feet to one side. A great shearing wave has passed through the town, moving roads and buildings to one side and creating a dislocation of its geometry. In an analogous way, when a plane-fronted wave of gravity, or curvature, passes through space-time, it leaves everything in front and everything behind perfectly flat, but across this wave things do not join up—there is a discontinuity or dislocation. Groups of null lines or light rays that approach the wave front become sheared and cannot join up.

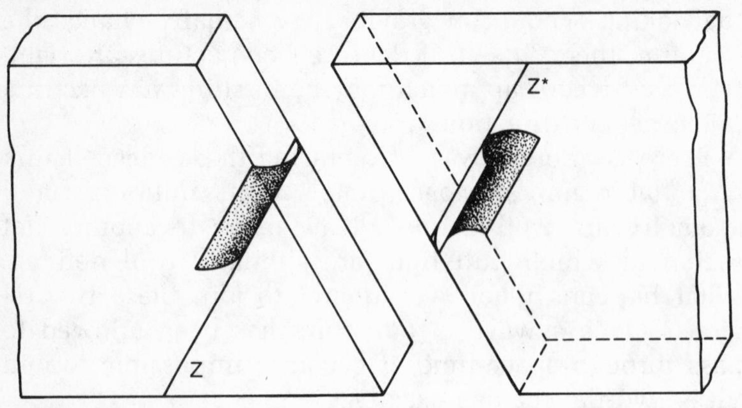

Figure 8–8
A plane-fronted gravitational wave passes through a previously flat space-time. Each of these two flat space-times now appears warped when viewed from the perspective of the other, and it becomes impossible to join them in a totally smooth way. The null line Z in one half of the space becomes the null line Z^* in the other. The corresponding picture in twistor space is for the twistor to become "mixed up."

The simplest form of space-time curvature could therefore be pictured as a failure of light rays or null lines to join up. But since null lines in space-time correspond to points in twistor space, the failure of the null lines to join looks, in the complementary twistor picture, like a relabeling of points in twistor space. Curvature in space-time therefore produces a transformation in twistor space. It turns out that this relabeling of points in twistor space is exactly what would be expected from a quantum mechanical transformation. So a gravitational wave or shear in space-time produces a quantum transform in twistor space. Conversely a quantum transform in twistor space will induce a shear wave or curvature in space-time.

Gravity and quantum process look like complementary pictures.

In fact, this picture, which links space-time curvature to a quantum transformation, while being exciting, turns out to be too simple. Penrose was to go on, a few years later, to write down the basic equation for a quantum graviton—the elementary quantum unit of curvature. Now, rather than patching space-time into local areas, twistor space itself would have to be patched. Penrose showed, for example, that within the global structure of twistor space, there is encoded information about the local curvature of space-time. While it turns out that large-scale curvature in space-time is the cooperative result of very many quantum gravitons, the effect of a single quantum graviton is too subtle to be discussed in such a crude way. To extend his approach, Penrose needed to create a new twistor picture for this single quantum of curvature. The nature of quantum gravity in the twistor picture is discussed in the next chapter.

Quantum Fluctuations

The twistor approach makes it possible to give a quantum theoretical picture of space-time and one that is very different from the conventional approach. The essential feature of the quantum theory is that it allows for probabilistic fluctuations of, for example, energy. But how exactly will these fluctuations affect space-time? The usual idea suggests that space-time becomes smeared out. In technical terms, its metric or basic measure fluctuates so that the distance between two points is no longer well defined. But this account posed a serious problem for Penrose, for it means that the light cone itself is affected, and this light cone, he argued, is the key structure in space-time. Once we allow for basic order of space-time, as expressed in its metric, to be subject to quantum fluctuations, then the light cone will become smeared out.

While other physicists were quite happy to demote the light cone from its privileged position in space-time, Penrose argued, for the sorts of reasons that were given in the previous section, that it was vital to preserve its basic structure. For Penrose, the spinor occupies a fundamental position (after all, a twistor can be described in terms of a pair of spinors), and therefore by implication the light cone must also be special. In other words, any theory that is built out of spinors must contain within it a clear conception of the direction of a light ray. Penrose therefore looked for a way of introducing quantum fluctuations into space-time while at the same time hanging on to the light cone. His answer was that the space-time point itself must become uncertain. Points are sacrificed while the global structure of the light cone remains unchanged.

The fact that space-time points are abandoned under the influence of quantum theory makes sense when we think of just how a point could be operationally determined. Points in space-time can be thought of as being defined through the interactions and collisions of quantum particles. But the elementary particles themselves are born in collisions and quantum interaction of other particles; they live out their lives until they either disintegrate or are swallowed up in another collision. An elementary particle is essentially a history in space-time, a world line. In this sense, it is a nonlocal thing having more in common with a twistor than with a point in space.

And how are points themselves pinned down? In practice they are the locations at which elementary particles collide, and are therefore defined by intersections of elementary particle world lines. Although in the past physics gave such significance to points in space, in an operational sense they are really secondary objects, derived from the intersection of global objects, the world lines of elementary particles.

This argument can be taken further. At low energies

the wavelike nature of an elementary particle causes it to spread out. To make a space-time point more precise, it is necessary to go to much higher-energy collisions. Yet in such collisions, many virtual particles are created, each with its own world line. In essence what we took for a point becomes an expression of the congruence of many world lines, which are all nonlocal objects. The "point" in a quantum mechanical description can never be exactly pinned down but must be thought of as fundamental, nonlocal, and with a smeared out appearance. This, as we shall see, is exactly the same answer we obtain in a formal, mathematical way from twistor theory.

In a theory in which points become secondary, smeared out objects, it may be possible to make quantum field theory calculations without running into infinities. (Many physicists believe that these infinities arise because space-time is assumed to be continuous so that physical quantities can be calculated right down to zero distances.) It may well turn out that infinities can be avoided in a different way from the superstring approach, this time by using twistors. At present this is still a research topic in twistor theory.

As we have already mentioned, the fact that space-time points are created out of the global structure of twistor space will be of great importance in discussing singularities. Where, in the conventional approach, space-time appears to break down at the core or singularity of a black hole, now this singularity becomes part of the global structure of twistor space—or at least this is one of the hopes of the twistor program. As Penrose puts it, while black hole singularities still exist, we may no longer have a "singular" space. The laws of physics can be defined over the whole of twistor space, and there will be no point at which they break down. It may also be possible to obtain a coherent account of the actual creation of the universe, that initial big bang in which everything that exists appeared out of an initial singularity.

It is also possible to revisit Wheeler's space-time foam. John Wheeler's argument was based on Heisenberg's uncertainty principle, which allows the energy within a tiny volume of space-time to fluctuate. When this region is as small as 10^{-33} cm, then the vast amounts of energy involved in these fluctuations act to break up space-time into a foam. But according to the twistor picture, that tiny region is also a global structure in twistor space. Possibly a new description is demanded at such short distances. The interpretation of high-energy scattering experiments suggests that, down to 10^{-14} cm, the conventional picture of space-time is probably correct. But what about shorter distances, distances that extend right down to that mysterious Wheeler limit? It may turn out that space-time has a radically different structure at these distances, one that demands an essentially quantum description. The first steps toward such a quantum space-time will be described in the next chapter.

Twistors and Particles

Lane Hughston has been an active member of Penrose's twistor group at Oxford. One of his interests has been to push ahead with the program of developing a picture of the elementary particles using twistors. Since the twistors themselves correspond to null lines and twisting congruences of null lines, they are essentially massless and create a conformally invariant picture in which no mass or sense of scale can appear. Objects that have mass must therefore be built up out of the interactions of twistors. (Remember that certain types of interaction break conformal invariance and introduce a sense of scale into space-time.)

The quantum fields corresponding to massive particles can be created out of the interactions of two or more massless twistors. And, since twistors themselves have their own built-in symmetries, the composite fields and particles generated will also manifest internal symmetries. It is the goal of the particle program that the known symmetries of the elementary particle families will be produced in this way.

The initial approach was to think of quark states as made out of three twistors. Since the twistors are created to have both linear and angular momentum, their composite states will also exhibit a range of angular momentum. Thinking back to Chapter 2 and the Regge trajectories, one of the essential problems was to account for high angular momentum of the elementary particle resonances. Now, it would appear, this could be done very naturally in a twistor picture. At first this picture seemed to be quite successful, but then the whole quark approach changed as the idea of "charm" was introduced. Suddenly the number of quarks jumped from three to six, and the corresponding twistor picture was inadequate. Hughston persisted, however, this time using six twistors as the basis for each particle. Penrose, for his part, felt that having to explain each elementary particle in terms of six twistors had become too artificial. For example, when the number of twistor constituents is extended from three to six, it becomes possible to build not only the known quarks but a vast number of other particles as well. Using six twistors as the ultimate building blocks of matter seems, to Penrose, to be too general.

Another active worker in this field of the twistor interpretation of elementary particles is the Hungarian physicist Zoltán Perjés, who began by writing down quantum wave functions in a complex twistor space. The quantum numbers that are related to the wave functions then have the appearance of some of the low-mass hadrons.

Another overwhelming problem, which has stumped all researchers in the elementary particle field, is the problem of mass. There are elementary particles like the electron, tau, and mu meson which appear to be identical in all respects except for differences in their mass. In several other cases, it is the mass that distinguishes one elementary particle from another. Working out the masses of the different elementary particles is a key problem in particle physics. However, as it stands, quantum theory does not tell physicists how to fix the values for mass. Indeed, when taken at face value, the theory predicts particles with every possible mass value. Likewise, twistor theory allows all mass values for the particles.

Penrose has always felt that solving the problem of mass can only be done by going outside quantum theory and including general relativity. After all, general relativity is a theory about mass, for it relates mass to space-time structure. It could well be that a proper understanding of mass will require fresh insights into the connections between general relativity and quantum theory. But this very field is one against which physicists have been banging their heads over the last few decades—the only result being severe headaches. Resolving the problem of mass and the elementary particles may have to wait for a much deeper physical theory.

Recently the twistor group has been taking another look at Regge trajectories. While this work is not immediately connected with the twistor approach, it may produce some interesting insights. Remember how, in Chapter 2, physicists had discovered that a graph of the angular momentum of elementary particles and resonances, when plotted against the square of their masses, produced a straight line. It was in an attempt to understand the meaning of these straight-line graphs, also called Regge trajectories, that the first string theory was born.

As new elementary particles were plotted, the picture became more complicated, and at the same time, the Regge approach went somewhat out of fashion. Recently, however, Penrose began to think about the meaning of angular momentum that is part of these Regge graphs. It turns out that all experiments designed to find out the angular momentum of an elementary particle end up measuring its square. So all we really know about an elementary particle is the square of its angular momentum, and not the actual angular momentum itself. Suppose, for example, we know that the number 4 is the square of some property. What is the actual value of this property? The answer is simple; the value must be 2 because $2 \times 2 = 4$. But, of course, there is another possible answer, the property could have the value -2, since $-2 \times -2 = 4$.

Penrose began to wonder if the same thing could happen with angular momentum. Suppose that for each measurement of a particle's angular momentum, there are two results—a positive and a negative angular momentum. Of course, the idea of a negative angular momentum sounds absurd. But just suppose that quantum theory admits of both negative and positive angular momenta. It is now possible to extend the idea of Regge graphs for both positive and negative angular momentum values. Sure enough, when this is done, the elementary particles and resonances do indeed appear to fall on a straight-line graph.

At present Sheung Tsun Tsou of the Oxford twistor group is looking at the experimental results on angular momentum measurements to see if the whole idea of negative angular momentum is really implied by the Regge plots. The group is also faced with interpreting the meaning of a negative angular momentum. Penrose, for example, wonders if this new approach could be connected to Nambu's first ideas on string theory so that hadrons could be interpreted in terms of vibrating and rotating strings that yield both positive and negative values for their angular momenta.

Twistor Networks and Twistor Diagrams

One of the initial goals in developing twistors was to extend the ideas of spin network in a fully relativistic way. With twistors in place of spinors, Penrose had hoped to create a fully relativistic space-time using only quantum mechanical rules. But twistor networks were to prove far more difficult than those based on spinors. In fact, until the new trouser diagrams came on the scene, Penrose was not actively working in this field, and the major research was being carried out by Andrew Hodges, who is also at the Mathematics Institute in Oxford.*

The first thing that Penrose and his colleagues had noticed about twistor networks is that the sort of pictures that can be drawn to represent the meeting of twistors are analogous to the Feynman diagrams used in conventional elementary particle physics. In other words, twistor diagrams express the way in which twistors interact. These interactions have the effect of deforming twistor space, and at the same time, they appear to introduce mass into the picture. Remember that conformal invariance of space-time is connected to the idea of massless particles, but this same conformal invariance can be destroyed by nonlinear interactions. The development of mass, the breaking of conformal invariance, and the appearance of certain forms of interactions are therefore all tied together.

In fact, it turns out to be very difficult to understand just what is going on. Even conventional elementary particle physics has problems when it comes to mass. The basic quantum mechanical rules give no reason as to why mass should be fixed at all. There seems to be no rule why the mass of the electron, for example, should not have a whole range of values. Moreover the various symmetry schemes that were created over the last two decades work best when the particle masses are zero. Mass seems to appear as a fine correction when these underlying symmetries are broken. Yet any comprehensive theory of the elementary particles will have to account for why

---

*Hodges' work on the twistor diagram approach was suspended for a number of years while he wrote and researched the highly recommended biography of the mathematician and computer theorist Alan Turing. Hodges, however, is now back at Oxford and vigorously pursuing this research.

particles have their own unique masses.

For the twistor theoreticians, the answer lies in general relativity. Since mass is the source of curvature in space-time, how will it ever be possible to understand its nature without bringing in both relativity and quantum theory? As far as the Oxford group is concerned, what is needed is some fresh understanding, an important new principle perhaps.

The spin network established the dimensionality of space. Beginning with binary objects, a space was created that referred to Euclidean angles in three dimensions. It was similarly hoped that twistor networks would uniquely determine the dimensionality of space-time as four. However, twistors are more general than this, and as abstract tools of mathematics, they have been generalized to higher dimensions, in particular to the ten dimensions of superstring theory. Nevertheless twistors do give some clues as to the dimensionality of our space-time.

It was earlier argued that it would be difficult for a universe to exist in fewer than three spatial dimensions, for then important topological interconnections would be impossible. Now Penrose has shown that the very important nonlinear constructions discussed in the next chapter are impossible in space-times with more than four dimensions. To have sufficient richness, our universe must exist in a four-dimensional space-time.

In addition, our world makes sense only when the time dimension enters as $-ict$. In calculating distances, for example, spatial terms make a positive contribution, while time makes a negative one. In this way it is possible to have null lines, light rays, whose length is zero. This can be expressed by saying that our space-time has a signature of $(+++-)$, indicating three positive spatial and one negative contribution. But it is mathematically possible to think up space-times with other signatures such as $(++++)$ or $(++--)$. These would give rise to totally different universes. It turns out that this particular signature of $(+++-)$ falls out of the twistor approach in a very natural way.

Finally the basic handedness of the twistors themselves and the way, as we shall see, they can divide a curved space-time into a right- and left-handed part indicates that twistors demand a chiral universe.

So although the twistor network approach is plagued by great difficulties, it does make it possible to introduce mass and elementary particle symmetries into the picture. In addition, twistors account for the dimensionality, signature, and chirality of nature.

Twistor diagrams and twistor networks are also connected with the idea of trying to make elementary particles out of twistors, a general program that was described in the previous section. Over the last two years, these programs had slowed down considerably, for the problems involved are extremely difficult, and what seemed to be missing was some essential new idea. However, a week or so after the manuscript for this book had been sent to the publisher, I received a letter from Roger Penrose telling me that there had just been what looked like a significant breakthrough in the general approach. My editor agreed to a short delay, and I went down to Oxford to visit the twistor group and discover what had happened.

The new idea appears to have originated with Michael Singer, although with a modesty that is characteristic of the twistor group, Singer claimed that it was just one of those things that come up in a discussion. It appears that the twistor group had been listening to a seminar on string theory and, while talking over tea with Roger Penrose and Andrew Hodges, Singer came up with the idea of extending the trouser diagrams of twistor theory by using twistor space.

Remember that the trouser diagram represents what happens when two string loops meet, merge into a single loop, and then separate into two loops again. Whole families of diagrams can be created simply by adding holes in the trousers.

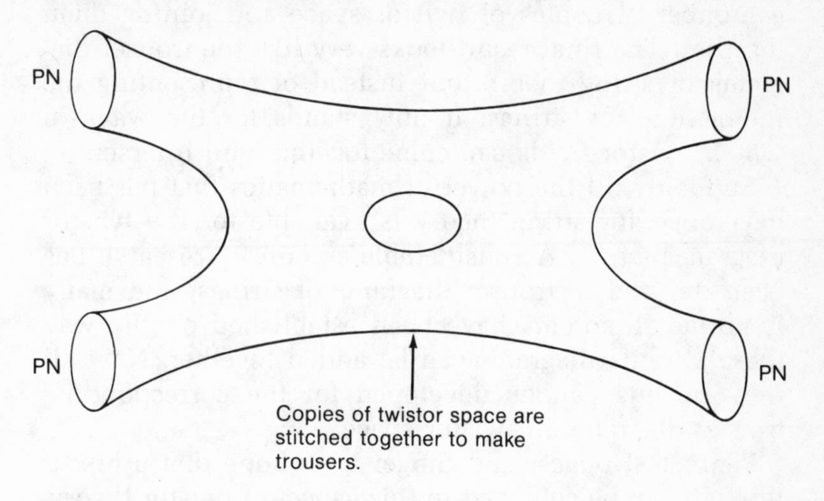
PN
PN
PN
PN
Copies of twistor space are stitched together to make trousers.

Figure 8–9
The trouser diagrams of string theory can be given a new perspective when viewed in twistor space. In string theory, they are the pictorial representation of how two free loops meet, interact, and emerge again as free loops. The complexity of the interaction corresponds to the number of holes within the trousers.

In the corresponding twistor picture, the free states are represented by the region PN of twistor space. The interacting region is created by stitching copies of twistor space together. Finally the free states PN emerge again.

Trouser diagrams begin with single-string loops. But what, Singer suggested, if we begin not with a loop but with a section of twistor space? The starting points are regions of *PN*—that part of twistor space whose points correspond to twistors of zero helicity, that is, to null lines or light rays in space-time. It is possible also to define the various massless fields of physics on *PN*.

Two regions of *PN* now represent the starting point in a new trouser diagram. They could correspond, for example, to two independent massless fields. Now extend these starting points or starting fields by building up the trousers. The trousers themselves are created by making a number of copies of twistor space and joining them together. The final result looks very like the trouser diagrams of string theory, but instead of representing the interactions of strings, it now stands for the ways in which twistor fields can come together and interact.

Suddenly all the powerful mathematics that has been developed for string theory is available for the twistor diagram picture. A considerable amount of research has been devoted to trouser diagrams of strings, and many important theorems have been established on the way these various diagrams can be added together. Now all these results can be developed for the corresponding trouser diagrams in twistor space.

Penrose, Hodges, and Singer now hope that a breakthrough can be achieved in this aspect of twistor theory. For example, it may be possible to develop new insights into the way mass appears in physics and how twistors and elementary particles are connected. Twistor trouser diagrams may give some new insights into how the masses of the elementary particles emerge. After all, trouser diagrams make connection with general curved space-time structure, since they are created in twistor space as well as with the quantum theory. The possibility of linking certain aspects of general relativity to the quantum theory may at some point hold the key to elementary particle masses.

While the diagrams in string and twistor spaces can call upon analogous forms of mathematics, it must be emphasized that the physical meanings of these two sets of diagrams are very different. On the one hand, string trousers represent closed loops that are "in" some sort of background space-time. The starting points of the twistor trousers, however, are not "in" any sort of space-time, rather they are regions of twistor space itself, which are then extended onto something that is analogous to an interacting region made out of copies of twistor space.

While the physical meanings of the two approaches are very different, it may be possible to draw on the many results of string theory in order to sort out some of the difficult problems involved in twistor diagrams.

Conclusions

Twistor space is the rich new arena of physics, for it is here, rather than in space-time, that the "great dream" must be explored. Already a complementary picture exists between the rich structure of twistor space and a space-time whose structure is built upon the global properties of null lines. In addition, the twistor approach suggests ways of relating gravitational curvature to quantum mechanical transformations. The next stage in the development of twistors is to create a new form of space-time, one with complex dimensions that will give meaning to the wave function of a single quantum of gravitational curvature. How this can be done, and the ingenious way in which the fields of nature are expressed using twistors, is the topic of the following chapter.

[illegible]

[illegible] experimental [illegible] [illegible] explored [illegible] [illegible] and [illegible] [illegible] [illegible] [illegible] [illegible] [illegible] [illegible] [illegible] and [illegible] [illegible] [illegible] the [illegible] [illegible] [illegible] [illegible] [illegible] [illegible] [illegible] [illegible] [illegible] [illegible] [illegible] [illegible] [illegible] [illegible] [illegible] [illegible]

9
Twistor Gravity

Twistor Fields

THE GREAT TRIUMPH of Penrose's twistor approach has been the elegant new way in which it describes the various fields used in physics. Fields have become one of the most important tools in modern physics. In the nineteenth century the electromagnetic field was created in order to explain the phenomena of light, electricity, and magnetism. Then, at the subatomic level, the idea of the field was to reappear as quantum field theory.

Take as an example Schrödinger's equation that describes the motion of an electron. This equation does not in fact explain the electron's origins or properties. Something more is needed. Quantum field theory, an extension of the quantum theory of Schrodinger and Heisenberg, attempts to go deeper. It begins with "classical" fields for matter and force and then goes on to quantize them. The quantum excitations of the electromagnetic field, for instance, become photons of light, while the quantum excitations of the electron field are electrons and positrons. The unified field theories of Chapter 4 begin with a single grand field whose basic symmetry is then broken. The quantum excitations of these symmetry-broken fields are approximations of the various hadron and lepton elementary particles.

The field description is fundamental in both classical and quantum physics, and it is here that twistors are able to provide a powerful new formulation—fields appear in a particularly natural way in the twistor space picture. But, since Penrose's approach is based on the proposition that mass is a secondary quality that arises in the interaction of more fundamental massless objects, the twistor formulation begins with massless fields such as those for the neutrino, photon, and graviton. With luck and some new insights, physicists may one day be able to discuss fields for massive particles within the same general formalism.

It turns out that these massless fields fall so naturally into the twistor scheme of things that it becomes possible to throw away the field equations themselves and discuss fields using a pictorial, geometrical approach! The twistor formulation of field theory is yet another step toward John Wheeler's "great dream."

Until Penrose and the twistor program came along, it was necessary to use what are called field equations in order to determine a field's behavior. But today, with the help of the rich cohomology (cohomology is a branch of topology concerned with joining together different regions of space—remember the illustration of the Penrose triangle) of twistor space, it becomes possible to get rid of the differential equations that determine the field. The twistor picture relies purely on the geometrical (or cohomological) properties of the field as it is expressed in terms of twistors and twistor space. This is a truly amazing result, for it means that the twistor approach can deal with the various fields of nature without ever needing to bother about field equations!

To understand just how this can be done, it is first necessary to know how things are traditionally calculated in physics. Think of trying to work out the path of Comet Halley as it moves toward the sun. Once astronomers have measured the position and speed of the comet, it becomes a fairly direct matter to use Newton's laws of motion to calculate the position of the comet at any future time. Physicists use these differential equations in order to calculate from point to point along the comet's path. Given an initial set of information—the comet's mass, speed, position, and direction at some initial time—it is possible to calculate the next point in its journey.

In the case of a field, this initial information must describe the whole of the field at a particular instant. It is like an instantaneous snapshot that gives information about the status of the field at every point in space. The field equations themselves are a set of differential equations that make it possible, given this initial snapshot, to calculate the exact state of the field at every point in space for some future time. So, with the help of the initial snapshot, the field equations generate a whole sequence of photographs, a movie if you like, of how the field changes in time.

This nineteenth-century image of a field in space must now be translated into the four-dimensional arena of space-time in which a relativistic field plays out its life. The initial information, or snapshot, can now be thought of as filling up a three-dimensional section of this space—a cross-section of space-time cut at one particular instant of time.

The field equations then act to extend this three-dimensional cross-section piecewise through the whole of four-dimensional space-time. The field equation could therefore be thought of as taking a three-dimensional surface, containing the structure of the field, and pushing this surface along the fourth dimension of time. In this way the whole of space-time is filled with the field. The field equation, being a differential equation, does this in a series of baby steps, so that an infinite series of three-dimensional cross-sections are put together in order to build up a continuous full four-dimensional space-time picture of the field.

The picture is radically different in Penrose's twistor approach, for the massless fields are now defined in (pro-

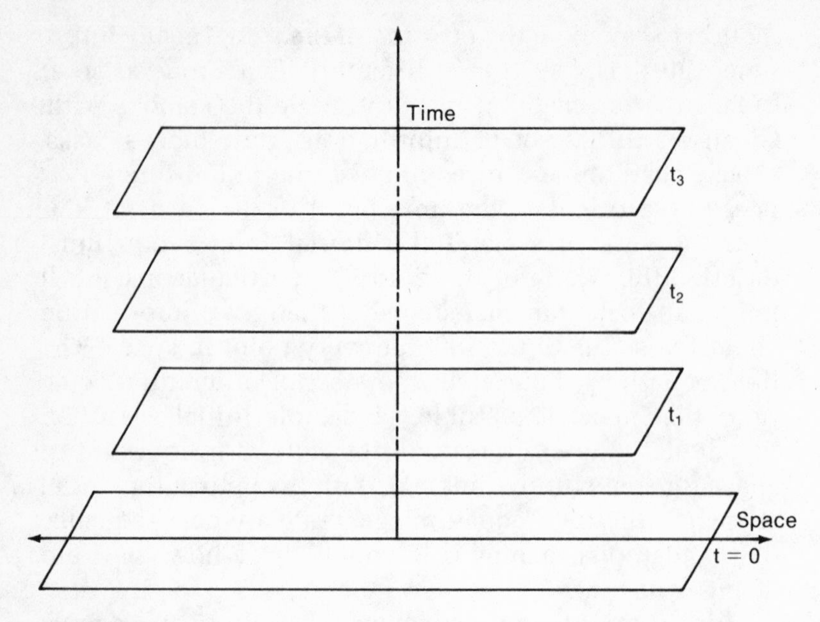

Figure 9–1
The state of a field is defined at successive instants of time.

jective) twistor space. But since this space has only three complex dimensions, it turns out that the information about the field's structure will totally fill twistor space! There is no room left in the twistor picture, nothing else for the field to do, no additional slice of space to fill! And, because twistor space is totally filled with the field's structure, there is no need for a field equation—the field along with all its dynamics is already totally defined, fully represented within the twistor picture. And, given the structure of this field in twistor space, it is then possible to recover the actual field along with all its dynamics in space-time. A point in the twistor space picture of the electromagnetic field, for example, becomes a ray of light in space-time.

Penrose was also able to show that each of the massless fields of nature can be created out of nothing more than a function of a single twistor. This is a truly remarkable result. It means that it is possible to write down a relatively simple mathematical function in twistor space that is so powerful it contains all the information that physicists need to know about the field at every point in space and for all time. In place of the differential field equations of nineteenth-century physics, Penrose has substituted a simple function in twistor space. The power of twistor mathematics is sufficient to define the field for all time and at all points in space.

The electromagnetic field begins as a fairly general function of a single twistor. It is the powerful mathematics of complex-dimensional twistor space that takes care of everything else and ensures that this function has the correct geometrical properties (or rather the generalization of geometry called cohomology). No longer must fields be described in terms of field equations, in which a differential equation moves the field from point to point. Bearing in mind the critique of the calculus and of differential equations that was made in Chapter 1, twistors have made a powerful leap forward.

The fact that the electromagnetic field, for example, can be created out of a general function of a single twistor would lead one to suppose that such a function must be very special and would be difficult to discover and write down. In fact, this is not true; all that is required is that the function should be mathematically well behaved. According to Penrose, the whole power of the twistor program is contained in what he would call the complex "analycity" of the twistor functions.

It is characteristic of such complex functions that they have points at which the function blows up—called singularities. In the twistor picture, the geometrical location of these singularities is of key importance; indeed the position of singularities in twistor space is related to the physical behavior of the field in space-time. The key to how this is done lies in an intuitively geometrical object called a *contour integral*. The contour integral is the essence of so much in the twistor approach; indeed, it enables us to calculate the way in which the twistor picture is changed by a single quantum of gravitational curvature.

Contour Integrals

Long before the Scottish physicist James Clerk Maxwell had worked out the field equations for electromagnetism in the mid-eighteenth century, Michael Faraday was picturing the phenomenon in terms of contour integrals. Of course, Faraday knew nothing about this particular branch of mathematics, but his great physical intuition had led him to see things in a particularly visual way. It later turned out that Faraday's ideas of "lines of force" that enter and leave magnetic poles and electrical charges can essentially be described in terms of contour integrals—the mathematics that Penrose was now applying to twistor space.

Suppose you scatter iron filings over a piece of paper placed on top of a magnet. The tiny particles of iron form a pattern on the paper, congregating near the north and south poles of the magnet, then stretching out as if to follow invisible lines of force that curve from one pole to the other. It was Faraday who pictured the phenomenon of magnetism in terms of fields of force that enter and leave magnetic poles. For example, the more dense the lines are in a given region of space, the stronger is the field.

Figure 9–2
The magnetic field of a bar magnet is most intense around its north and south poles. The field lines are correspondingly drawn closer together in these regions.

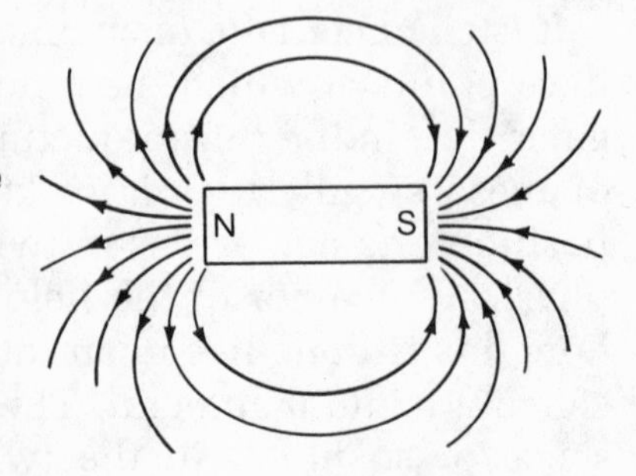

Exactly the same explanation can be made for an electrical field—lines of electrical force enter positive electrical charges and leave the negative charges. Faraday had hit upon a highly intuitive and visual way of describing electrical and magnetic fields in terms of lines of force. Indeed, the number of these lines in any given region would be equal to the strength of the field.

Faraday's approach was based upon the ideas of boundaries and the number of lines that pass through a given surface in space—ideas that occur, in a more sophisticated form, in cohomology theory. In the following diagrams, one simply draws a boundary around a given area and counts the number of lines inside. The greater the number of lines that pass through a given area, the greater is the strength of the field.

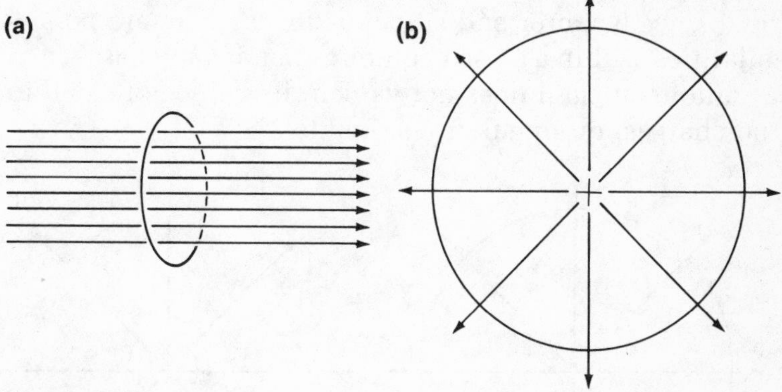

Figure 9–3
(a) Since there is no charge within this region, the same number of fields enter and leave.
(b) Field lines originating on a positive charge cross the boundary.

Faraday's idea of drawing a boundary or contour around the electrical and magnetic charges anticipates the essential features of modern contour integration. Look, for example, at the closed contour below. Because there is no electrical charge inside, the same number of lines enter as leave the contour. In terms of contour integration, this would mean that the value of the contour integral is zero. (The word *integral* means a summing up.)

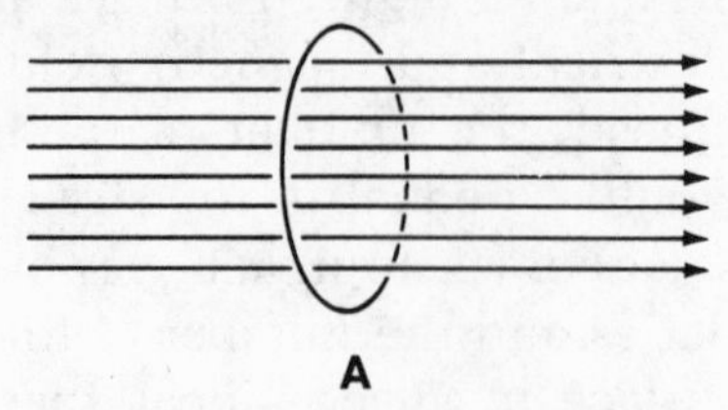

Figure 9–4
Since the same number of lines enter as leave, this contour integral is zero.

In the next set of diagrams, there is an electrical charge inside the contour. This means that more lines leave the contour than enter, and the contour integral no longer vanishes. Its value is equal to the size of the charge inside the contour. The charges act as sources for the electromagnetic field, and mathematicians call such places where field lines begin or end *singularities*. If there are no singularities within a closed contour, then it vanishes. When a contour integral does not vanish, its value is related to the charges, or singularities, inside.

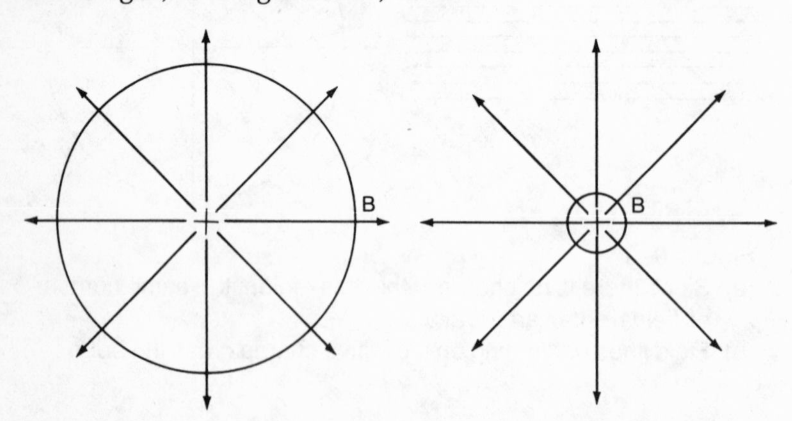

Figure 9–5
The value of the contour integral is given by the number of field lines that cross its surface. Even when the contour is shrunk, the same number of field lines cross its surface. Clearly the contour can be shrunk right down without changing its value, provided that it always surrounds the charge.

What is important about these contour integrals is that it is possible to deform them without changing their value. In the case of the contours *A*, it doesn't matter how big or small they are made, the integral is still zero, for the same number of lines enter as leave. In *B* we can shrink the contour right down, and provided that it still goes around the charge, or singularity, it will have the same value. In *C* we can distort the contour so that it just goes around the two singularities.

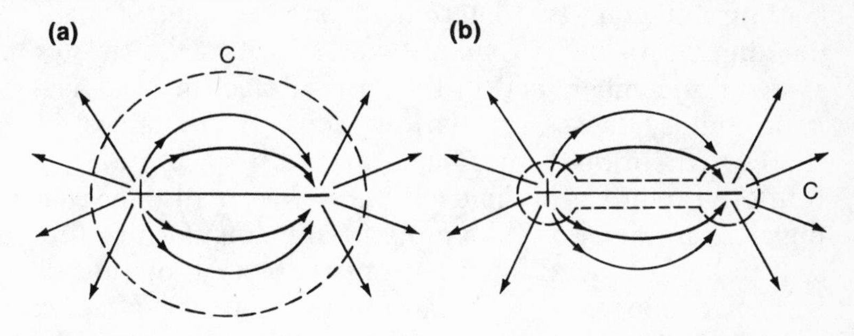

Figure 9–6
The contour *C* can be shrunk until it surrounds the two charges, or field singularities, which are sources and sinks for the field lines.

The value of a contour integral does not depend upon its particular shape but what mathematicians would call its cohomological properties alone. In other words, the shape of a contour integral can be distorted without changing its value; all that matters is its property of being a boundary around the charges or singularities. In this way electromagnetism depends upon a branch of cohomology (called by mathematicians "sheaf cohomology"). It turns out that this same cohomology forms the backbone of much of Penrose's twistor work—and recently is also being used at the frontiers of superstring theory.

Twistors and Contours

Now back to Penrose's twistor picture in which the massless fields of physics can be defined within twistor space without the need for any field equations. We have already seen that the definition of a massless field depends upon certain general functions of the twistors. It turns out that the only restriction upon the fields is a very beautiful mathematical one: *Any massless field is defined by a contour integral in twistor space.* And these contour integrals are determined once we know about the singularities of a general twistor function in twistor space. (Remember that, in the case of electrical or magnetic fields, these singularities look like charges or sources where field lines begin or end.)

In other words, the fields of nature, with all their rich physics, are determined simply by the singularities present in fairly general functions of twistors. All of physics has been reduced to the beautiful geometry of complex spaces. Clifford's great dream has been partly realized.

This is a staggering result. In place of field equations—differential equations that determine how the field changes in time—we have a simple mathematical function, one whose analytical structure is sufficient to determine all the possible behaviors of, for example, light and the electrical and magnetic fields. The field is given by a contour integral around this function, and this contour integral is simply determined by the nature of the singularities of this function.

Of course, once the field is known in twistor space, it is then possible to re-create the field in its corresponding space-time picture, for the twistor and space-time pictures are complementary. For example, the points in twistor space correspond to global properties of the electromagnetic field in space-time such as light rays or congruences of light rays. Armed with a general function of a single twistor, it is therefore possible to reproduce the electromagnetic field, along with all its rich dynamics, in space-time. Indeed, this can be done for any massless field in nature.

When I first met Roger Penrose seventeen years ago, he was convinced that nature must be determined in a combinatoric way using only the natural numbers. Today, as he puts it, his position has mellowed. While counting may be of primary importance, it nevertheless conceals beneath it a rich world of complex geometry. It is this rich structure that has taken twistors into so many new fields of mathematics and theoretical physics. Today twistors are being used not only to determine the structure of space-time and the elementary particles but in ways Penrose never anticipated, such as in the solution of difficult nonlinear differential equations and a particular technical problem known as the Yang-Mills instantons.

The Photon Wave Function and the Electromagnetic Field

The electromagnetic field, in all its richness, does not require twistor field equations, for it is simply defined in terms of a contour integral taken around a twistor function. Once the twistor picture has been established, it is then possible to re-create the complementary space-time description of the electromagnetic field. This same procedure can be carried out for all the massless fields of nature. It is also hoped that one day it will be possible to take account of the origin of mass and then describe all of nature's fields in a twistorial way.

But now let us look a little more closely at this function of twistors that can reproduce such fields as electromagnetism, gravity, or neutrinos, this function around which a contour integral is to be taken. It is important to ask what restrictions are placed on it by the actual physics of a particular field. In other words, what distinguishes the twistor function for the photon from that of the neutrino or graviton?

The actual equations that James Clerk Maxwell wrote down for the electromagnetic field explain, among other things, how light and radio waves can travel through empty space. Suppose that oscillations are set up in an electrical field. This changing electrical field now acts to generate an oscillating magnetic field, which is located at right angles to the electrical field. In turn, the variations in the magnetic field are the source of oscillations in the electrical field. The whole thing therefore works like a feedback loop. It is self-sustaining because electrical oscillations produce magnetic oscillations, which in turn create electrical oscillations again. The whole disturbance moves through the vacuum of space at the speed of light and is experienced as light or radio waves.

It turns out that Maxwell's expression for the electromagnetic field demands two indices, and the quantum particles or photons corresponding to this field have a helicity of either +1 or −1. This helicity can be translated directly into what is called the homogeneity of the twistor function, and this homogeneity is sufficient to tie down the twistor function. The power of the twistor picture is therefore amazingly simple.

The homogeneity of a function could be thought of as a count of the number of powers it contains. For example, a function like $x^2y + y^2x + x^3 + y^3$ has a homogeneity of +3. Terms like x^5 and x^2y^3 are associated with a homogeneity of +5. It is also possible to have negative homogeneity with general terms like $\frac{1}{x^4}$ for a homogeneity of −4.

In the case of a field whose quantum particles have a helicity of +1, it is necessary to write down a twistor function of homogeneity of −4. One simply writes down the most general twistor function having this homogeneity (that is, containing powers like $\frac{1}{x^4}$) and then wraps a contour integral around it and looks for the singularities inside. The result is a complete definition of a photon of light in twistor space!

To sum up again, Penrose's approach was to get rid of differential equations and express the electromagnetic field purely in terms of what mathematicians call a sheaf cohomology. Sheaf cohomology is the general geometrical way of dealing with contour integrals. It has to do with the ability to cover a complex space with overlapping regions that are smooth and well behaved. The sheaf part of this cohomology allows us to wiggle patches of space around in order to make them fit together. Differential equations are therefore thrown away, and the twistor picture of a field is now given cohomologically in terms of the way a contour wraps itself around a particular function in twistor space.

The Left and Right Hand of Light

Now let's look at the electromagnetic field in a little more detail, for it turns out that the twistor picture shows something exciting about the quantum nature of light. Light, the ordinary light around us, can really be split into two parts called planes of polarization—these could be thought of as right- and left-handed planes. This is easily seen with polarized sunglasses.

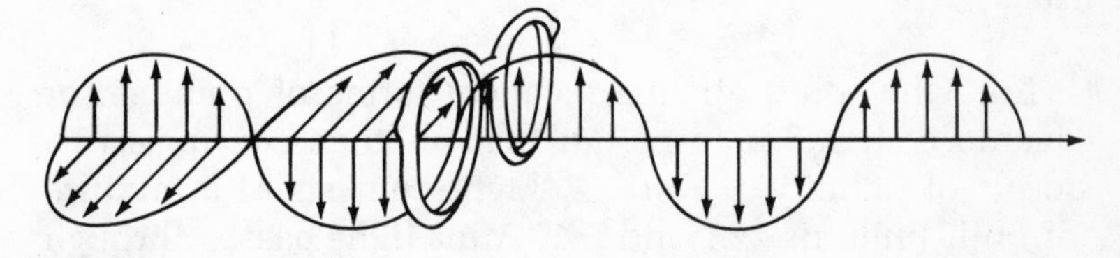

Figure 9–7
The electromagnetic field corresponding to normal light vibrates in two polarized forms. Here light passes through a pair of Polaroid sunglasses, which filters out one set of these vibrations.

The polarizer in the lens lets in light that is polarized only in one of two directions. In fact, the geometry of the sunglasses cuts out *glare*—that polarized component of light which is scattered or bounced back into the eyes by shiny objects such as windows, water, or metal. This phenomenon is easy to see by taking a second pair of polarized sunglasses and holding them in front of the first. Provided they are both aligned, light still passes through—they are both filtering out one of the two polarized components of light and letting the other through—but if one pair of sunglasses is rotated to be at right angles to the first, the second polarized component is also removed. With both components removed, the result is blackness.

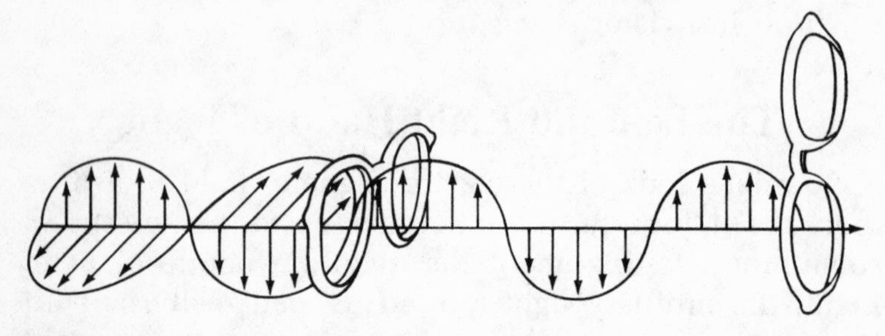

Figure 9–8
A second pair of sunglasses, oriented at right angles to the first, filters out the second set of vibrations so that no light is transmitted.

But what does all this mean in terms of photons, or quanta of light? It turns out that photons can have two forms of helicity, +1 or −1. Light consists of a mixture of both photons, +1 and −1. After light passes through the polarized sunglasses, only +1 photons, say, remain. But if the second pair of glasses is then rotated at right angles, they will also filter out the +1 photons, leaving blackness.

It is at this point that the twistor picture becomes really exciting. As we have seen, photons with helicity +1 are

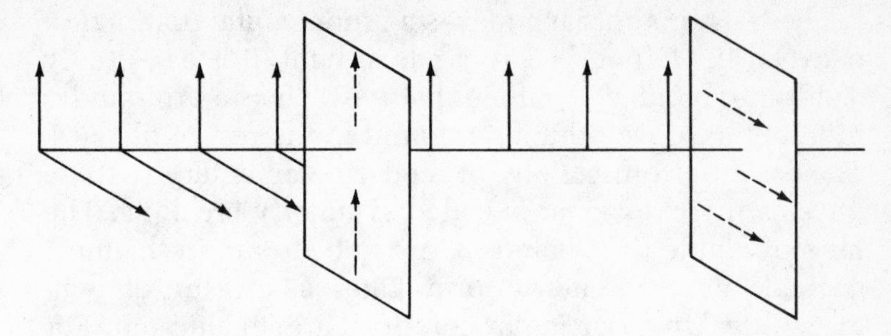

Figure 9–9
Individual photons are represented quantum mechanically as a linear combination of two helicity states. The action of a polarizer is to "measure" each photon, that is, to force it into one or another of the states. Only 50 percent of the photons pass through the first polarizer.

Those photons which have passed through the first polarizer are now all in the helicity state. Orienting the second polarizer at right angles to the first filters out all these photons. While the classical and quantum mechanical explanations may be different, the results are the same: no light will penetrate a system of crossed polarizers.

represented by a contour integral around a twistor function of homogeneity -4 (that is, a function containing terms like $\frac{1}{x^2y^2}$). But in the case of a photon with helicity -1, a twistor function of homogeneity 0 is needed.

Now, this is a particularly surprising result, for it suggests that a proper description of a general beam of light involves two very different twistor functions. It is not simply a matter of one function being a reflection or mirror image of the other; these are totally different functions—one with homogeneity -4, the other with homogeneity 0. Clearly, within the twistor picture of things, light is essentially asymmetric in its most fundamental description. Its basic law does not have an easy symmetry between $+1$ and -1 helicities; rather the two polarized forms of light are generated in totally different ways using twistor functions of completely different helicity.

The twistor approach makes a fundamental distinction between the left and right hands of light. It is essentially a chiral or handed picture of nature. This is a profoundly different approach than that found in the rest of physics. The essential philosophy of modern elementary particle physics involves what is called symmetry breaking. The laws of nature, it is believed, are to be found in the most general, most symmetric form. Later, as a result of symmetry breaking, particular solutions are singled out. For example, the universe is created in a totally symmetric way, with handedness or chirality later emerging as a form of symmetry breaking. But Penrose is saying the direct opposite, that in their most fundamental form, the laws of nature are *not* symmetric but exhibit a basic handedness or chirality.

To sum up, the right-handed photon is given by a contour integral in twistor space taken around a twistor function of homogeneity −4, the left-handed photon requires a contour integral around a function of homogeneity 0. Left- and right-handed photons emerge in totally different ways.

But there is yet one more turn of the screw. Quantum theory teaches us that any linear combination of quantum solutions must correspond to a physical solution. In other words, quantum theory dictates that single photons with mixed helicity—combinations of left- and right-handed photons—are perfectly valid. Within the twistor picture, this means that a photon can be generated out of a two-twistor function, each having a different helicity. The twistor approach is clearly giving us a new way of thinking about the wave function of the photon.

This approach of representing massless fields by contour integrals of twistor functions is, of course, perfectly general and can be applied to cases other than the electromagnetic field. The following table gives the homogeneity of the twistor functions that generate each particular field. In each case a right- and a left-handed quantum particle are represented by a function of a totally different homogeneity. At its most fundamental nature, this twistor picture is not symmetric.

Particle	**Homogeneity**
graviton +2	−6
photon +1	−4
antineutrino $+\frac{1}{2}$	−3
0*	−2
neutrino $-\frac{1}{2}$	−1
photon −1	0
graviton −2	+2

An interesting implication of the table is that the property of a particular field, or quantum particle, being a fermion (with fractional helicity or spin like $\frac{1}{2}$) or a boson (with a whole number helicity like 1 or 2) now depends upon the homogeneity of functions in twistor space. This is a particularly suggestive idea, for it indicates that there must be a fundamental interconnection between the division of elementary particles into fermions and bosons and the underlying structure of twistor space.

The next step in the twistor program would be to generalize the contour integral approach to massive fields and in this way attempt to generate the known elementary particles as quantum excitations of these fields. Mass, it appears, is created out of the interaction of twistors and means that massive fields have to be generated using contour integrals around functions of more than one twistor. This is a difficult and complicated problem whose implications are yet to be worked out. One possibility lies in the recent idea that the trouser diagrams of superstrings can be reinterpreted in twistorial terms. It

*The spin 0 particle whose twistor function has homogeneity −2 in the table does not correspond to any of the known particles of physics. Its possible physical meaning remains an open question.

is possible that using this new approach will shed some new light on the nature of twistor interactions.

Marble and Straw

One of the most exciting developments in twistor theory has been the way in which Penrose extended the twistor picture of the massless field to include gravity. In fact, it becomes possible to write down the wave function for a graviton and give a meaning to a single quantum of gravitational curvature.

It turns out, in fact, that within the twistor picture, a single graviton requires a very different picture than that for a single photon. Indeed, twistor gravity brings us right to the heart of the paradoxical division between quantum theory and general relativity.

In the previous section, Penrose had discovered how to express the wave function for a photon in terms of the rich properties of twistor space. Why is it that this approach cannot be repeated in the case of a graviton? The reason is that a graviton is not only the quantum particle of the gravitational field, it is also a quantum element of the curvature of space-time itself. Gravity is both a force that can be measured between two massive bodies and a theory about the curvature of space-time. In other words, admitting a single graviton into the twistor picture actually changes its geometry. The beautiful picture of massless fields in terms of contour integrals around twistor functions depends critically upon the powerful mathematical properties of twistor space—but the appearance of a single graviton will transform these properties. Paradoxically the very power of twistor space to describe gravity is disturbed by the existence of what it had set out to define!

How Penrose was able to extend his twistor picture into the domain of quantum gravity will be described in the following sections. But in order to understand what

Penrose did, we must first go back to Einstein's great discovery.

Albert Einstein's general theory of relativity explains how the curvature of space-time affects the motion of matter and how this effect is experienced as the force of gravity. In turn, the theory must also explain how the presence of matter and energy gives rise to this space-time curvature.

$$R_{\mu\eta} - \frac{1}{2}g_{\mu\eta}R = -kT_{\mu\eta}$$

Figure 9–10

When Einstein had created his general theory of relativity and written down his field equations, he is supposed to have said that while the left-hand side had been carved in marble, the right-hand side was built out of straw. The left hand of Einstein's famous field equations refers to the actual geometry of space-time and is one of the great insights of science. The right hand describes how matter and energy produce this curvature, but this description did not follow with such elegant inevitability as the other side. Only after Einstein had written down the part of his theory describing the actual geometry of space-time did he turn to the assumptions necessary to create this geometry—the part of the theory that is "built out of straw."

One of the clues that helped Einstein to carve his theory in marble came from asking why the gravitational mass of a body is always equal to its inertial mass. The usual way of discovering the weight of a body is to place it on a scale or spring balance. In fact, what is being measured is the amount of gravitational pull the earth exerts on the body. When you stand on the scale in the morning, you are really learning about the force that is mutually exerted between yourself and the earth. The amount of this force is governed by what is called your *gravitational mass*.

But there is another way of discovering how massive something is, and that is by pushing it. By experiencing directly just how difficult it is to get a body moving and how much force is needed to slow it down, we also discover how much "stuff" it contains. Going on vacation with the whole family and their luggage, you will find how hard it is to accelerate the car when overtaking on the highway and that a longer safety factor is required when braking. The more a body resists a change in its motion, the more mass it contains. Physicists call this form of mass its *inertial mass*.

Until Einstein came along, there was no theoretical reason why the gravitational and inertial masses should be equal. After all, one of them has to do with the resistance to changes of movement, and the other has to do with the degree to which the body is pulled by gravity. It could turn out, for example, that the ratio between these two masses could depend on how they were constructed or what they were made out of.

Suppose that the ratio between gravitational and inertial mass was a function of the composition of the bodies and was therefore different for iron and stone. Let us follow Galileo up the leaning tower of Pisa while he simultaneously drops a ball of iron and a ball of stone from the top of the tower. The greater the gravitational mass, the greater will be the force pulling these bodies to the ground. On the other hand, the greater the inertial mass, the more the bodies resist this acceleration. It is not difficult to see that if the ratio between gravitational and inertial mass varies between bodies, they will all fall at different rates. But Galileo discovered that, neglecting air resistance, all falling bodies hit the ground at exactly the same time. At the beginning of our own century, the Hungarian scientist Roland von Eötvös determined this equivalence to be 1 part in 10^8, and more recently it has been confirmed to be 1 part in 10^{11}. But why should these two forms of mass be equal? Is it a matter of coincidence, or could there be some deeper reason?

It was Einstein who suggested that the solution to this problem could be resolved if the motion of bodies and the force of gravity are understood purely in geometrical terms. Einstein argued that space-time is not flat but curved and that, in their movement, bodies take special paths called geodesics through this curved space-time. Motion is determined purely on geometrical grounds. Since the space-time curvature generated by the earth will be the same for all small bodies, they will all move along common geodesics in space-time. That is, they will all fall at the same rate through space. This geometrical interpretation of space-time and motion will only correspond to experience if the inertial and gravitational masses are equal. Their observed equality is an inevitable consequence of the geometrical interpretation of gravity and motion.

It was such insights into the meaning of motion and gravity that enabled Einstein to create the left-hand side of his field equations. These describe, among other things, the paths taken by small test particles such as stones and cannonballs. (The test particles must be small enough not to exert an appreciable effect upon the geometry of space-time themselves.) All physicists agree upon the mathematical beauty of this part of Einstein's theory and the way it follows so naturally from Einstein's basic insights.

But Einstein still had the problem of describing how the geometry of space-time is itself determined by the presence of matter and energy. Geometry governs the motion of matter and, in turn, matter determines the geometry of space-time. Where in the prerelativistic world, mass had entered physics as a measure of resistance to change of motion (inertial mass) or to the force with which a body is pulled to earth (gravitational mass), it now appears in a totally new way, on the right-hand side of Einstein's equations, as the source of space-time geometry. To complete his theory, Einstein had to show how various configurations of matter and energy will give rise to the specific geometry of space-time.

While the left-hand side of Einstein's field equation describes the geometry of space-time, the right-hand side has to do with matter and energy. Even at this starting point, one feels uneasy; what on earth does it mean to equate geometry to matter and energy? No wonder Einstein dreamed of a unified theory in which matter and energy could themselves be reduced to the hills and knots of space-time. In such a theory the right-hand side would vanish, and everything would be described in terms of the side carved in marble, the geometry of space-time.

But neither Einstein nor any of his contemporaries knew how to write down such a unified theory, and he was therefore obliged to go along with this equating of geometry with the material part of nature. He was therefore forced to make a number of what appear to be very reasonable assumptions in order to create the right-hand side of the field equations, which express how matter acts as the source for space-time geometry. The left- and right-hand sides of the combined field equations reduce to the Newtonian account of gravity in the limit of slow speeds and weak gravitational fields. In addition they describe the small deviations from Newtonian law that have been observed when very accurate measurements are made on the orbits of satellites as well as such phenomena as the bending of starlight as it passes through the sun's gravitational field and the slowing down of atomic clocks in a gravitational field. General relativity has been well tested, yet Einstein was never really satisfied with the final form of his great theory, for some much deeper principle was required to determine the right-hand side of his field equations.

Today we realize that matter and energy cannot be understood without the help of the quantum theory. But this sheds a new light upon Einstein's field equations. The right-hand side contains a description of matter that is purely classical, yet we know matter and energy to have a quantum mechanical origin. One could say that since the right-hand side of the equation is about matter and energy, it must ultimately be quantum mechanical. Yet the left-hand side is about space-time, which has hitherto resisted all attempts to be quantized. Conventional wisdom has it that both sides of the equation must be quantized. Or rather, Einstein's field equations must be derived from some more fundamental theory which talks about geometry and matter at the quantum mechanical level. Such an approach would yield a new type of space-time structure appropriate to wave functions of individual gravitons. Once such a formalism has been developed, it should then be possible to derive Einstein's field equations as the large-scale limit. Just as Newton's equations earlier were recovered in the limit of weak fields and low speeds, so too Einstein's theory would also become an approximation.

Quantizing Gravity

The conventional approaches to quantizing gravity and space-time work in the following way. Quantum processes occur only on the smallest scale and are almost undetectable at the human level. It therefore seems reasonable, the argument goes, to assume that an individual quantum of curvature has an almost negligible effect on the geometry of space-time. Only when many gravitons are present should space become noticeably curved.

The general approach is to begin with a flat, empty, Minkowski space-time and add on gravitons. Mathematically this amounts to a perturbation process. We have met the idea of perturbations several times before in this book. It involves an infinite number of very tiny corrections that add up to produce a finite effect. Einstein's equations are essentially nonlinear so that gravity feeds back into itself (the idea of nonlinearity will be explored in another of the author's books, *The Turbulent Mirror,* coauthored with John Briggs and published by Harper & Row.) But quantum physicists don't like working with nonlinear theories, so they begin with a linear, flat space-time approximation of Einstein's theory. The basic idea is that when the infinite number of corrections or perturbations have been added, the correct result will be established.

There is yet another assumption involved in the conventional approach, and this is that conventional quantum theory is sufficient to discuss the quantization of general relativity. In other words, the curvature of space-time and the essential nonlinearities of Einstein's theory will have no effect on the quantum theory itself. Quantum theory, it is assumed, is prior to general relativity and space-time structure.

The problem with this whole approach is that it doesn't seem to work. Space-time begins totally flat; add a single graviton, and it remains flat; add a handful, a bucketful of gravitons, and it is still flat—the theory remains linear. In fact, space-time remains flat until an infinite number of gravitons have been added to the space. This result doesn't seem to make much sense. How can it be that a large collection of gravitons have no effect on the structure of space-time, yet in the limit of an infinite number of gravitons, the space-time suddenly becomes curved?

In essence, this result is really showing up the limitations of perturbation theory and the strange things that can happen when an infinite series is added. A simple illustration may help. Take the function $\frac{1}{x}$. For a large x this function is small, but for a small x it increases. In fact, when $x = 0$ the term $\frac{1}{x}$ blows up to infinity. The same thing happens with the functions $\frac{1}{x^2}$, $\frac{1}{x^3}$, $\frac{1}{x^4}$, and so on. In each case, the function is very small for a large x but blows up to infinity when $x = 0$.

Now take the sum:

$$\frac{1}{x} + \frac{1}{x^2} + \frac{1}{x^3} + \frac{1}{x^4} + \cdots + \frac{1}{x^n}$$

where n is some number. This series becomes infinite

when $x = 0$, no matter whether $n = 10$ or 100 or 1,000.

But now suppose we let this series go right on to infinity. Something very curious happens, for it is mathematically possible to express this infinite series in a closed form as:

$$\frac{1}{x-1} = \frac{1}{x} + \frac{1}{x^2} + \frac{1}{x^3} + \frac{1}{x^4} + \ldots + \frac{1}{x^n} + \ldots \infty$$

This series no longer blows up at $x = 0$; in fact, it equals a finite number, -1. The series does, however, blow up when $x = +1$.

This is one of those strange occurrences that can happen in mathematics. For finite n, the series goes along quite happily, blowing up when $x = 0$, right until the series becomes infinite, at which stage the point of blowing up shifts from $x = 0$ to $x = 1$! This can be taken as an illustration of the limits of the philosophy of perturbation theory, for it means that by adding progressively more and more corrections, we can always think that we are homing in on the right answer, but as soon as the series becomes infinite, the answer changes.

Penrose felt that something similar was happening with the quantum perturbation approach to gravity. He could not accept that space-time remains flat until the last straw is added and the number of gravitons increased to infinity. Rather, he argued, "each graviton must carry its own measure of curvature." But what is the meaning of a single graviton? The question has a Zen feel about it. A single graviton must be a very strange object, a quantum element of curvature, something that is both quantum mechanical and relativistic, yet that lies beyond the accepted conventions of either theory. Clearly physics had to go beyond the conventional space-time picture in order to understand a single graviton. Space-time is too gross and too simple. What is needed is a much richer picture, a complex space-time.

Classical gravity is pictured in conventional space-time as a curvature of geometry. But in the quantum picture, gravity and space-time curvature arise through the actions of gravitons: quantum mechanical objects described by wave functions. Now Penrose was arguing that single gravitons cannot be defined in space-time but require a new complex space—nothing less than the full space of twistor theory itself. Earlier in this chapter, we learned how a twistor space picture of massless fields could be created using a contour integral in twistor space. The approach works because of the rich geometrical properties of twistor space. But now the idea is to go from photons to gravitons and express the first elements of quantum curvature in terms of the twistor picture. Yet the first appearance of curvature is enough to disturb all this beautiful mathematics of twistor space. How was Penrose to proceed?

It became clear that the complementary pictures of space-time and twistor space were simply not general enough to encompass the new world of quantum gravity. What was needed was a much richer space with more general properties. Now it turns out that the "twistor space" *PT* we have been using up to now is not really the most general form of space for twistors. In fact, *PT* is a projective space, that is, a reduction in three complex dimensions of the full four-complex-dimensional twistor space *T*. While this projective twistor space *PT* contains enough structure to describe the wave function of a photon or neutrino, the full force of the more general twistor space *T* is needed when it comes to describing gravity.

Remember how, in Chapter 8, a complementarity was established between the space-time and projective twistor pictures—points in twistor space correspond to lines or congruences of lines in space-time and vice versa. Now a complementarity is established between full twistor space and a new complex space-time. (This complex space-time is the complex generalization of the space-time in which we live.) As before, there is a duality between lines in one picture and points in another. As before, there is a special region that bisects the space so that there is a physical meaning to negative- and positive-frequency solutions. The new picture preserves all the beautiful structure we learned about in the two previous chapters; it is simply more general.

It is, of course, possible to duplicate everything that has been done up to now, but this time in full twistor space and complex space-time. For example, the wave function for a photon or neutrino can be generated in twistor space, using a contour integral, and then the corresponding (complex) space-time picture recovered.

At the moment this space-time is flat and is therefore not yet able to express the meaning of a graviton of curvature. In Chapter 8 the appearance of space-time curvature was equated with a transformation among the points of twistor space. This was interpreted as producing actual quantum mechanical changes. Conversely a quantum mechanical transformation in twistor space produced a wave of curvature in space-time. But this explanation does not really go far enough; it is a simple picture that requires a more rigorous mathematical support. Now, using the contour integration approach, it becomes possible to discuss the meaning of quantum curvature in a much deeper way.

The appearance of a graviton in full twistor space transforms the mathematical properties of that space, for while the graviton preserves the local structures, it produces global changes in twistor space. But it is the global structure of twistor space that determines the points of space-time—global structures in twistor space determine local structures in space-time. In the full twistor picture this means that an individual graviton smears out the points of our complex space-time.

To find out what happens in greater detail, we must return to the contour integral approach. As before, the graviton field splits into two parts: a helicity +2 graviton and a helicity −2 graviton. The field, or wave function, for the helicity +2 graviton is generated using a contour integral that must be wrapped around the twistor function with a homogeneity of −6. To obtain a graviton of helicity −2 it will be necessary to use a twistor function with homogeneity +2. As before, the field is determined by sliding the contour integral around the function and making sure that all the singularities are enclosed.

The appearance of nonlinear gravity in full twistor space distorts it and the complex space-time. One way of dealing with a curved and distorted space is to try to cover it with small flat patches. But now, rather than patching space-time, it is the full twistor space that must be attended to, for it has become distorted by the full power of nonlinear gravitation. (In this approach only two patches are in fact needed.)

The individual flat patches of the contour integral now have to be moved around, and this means that the structure of twistor space will differ from region to region (since these flat regions are attempts to cover the distorted structure of twistor space). Straight lines will no longer join in twistor space once its global structure has become deformed. But the global structure of lines in twistor space is what gives meaning to points in complex space-time. Does this mean that space-time no longer has meaning?

It turns out that while straight lines no longer join up, they can be replaced by very general curves that have definite structures. These are called holomorphic curves, and their structure is unchanged under twistor transformations. (We shall meet these same holomorphic curves again, when we shall discover that a holomorphic curve in twistor space corresponds to a superstring in space-time!) The structure of these general curves is still preserved even with the addition of gravity into the twistor theory, while straight lines are, of course, destroyed by this curvature. So all is not lost when quantum curvature appears, for the important structure of very general curves is unchanged.

Working in full twistor space, it is possible to extend

the basic contour integral approach to cases where the quantum particles actually disturb the nature of twistor space itself. While quantum gravity will transform the global structure of twistor space so that straight lines can no longer be clearly defined, it is still possible to work with more general curves. And since the structure of these lines is unchanged by quantum gravity, it means that key parts of the powerful twistor mathematics are still available. Even though gravity disturbs the global structure of twistor space, it is still possible to draw upon the complementary twistor/space-time pictures.

But what meaning does all this have in the corresponding complex space-time? It turns out that while local things like space-time points are no longer well defined, global things like null lines and light rays still have meaning. While space-time may be distorted locally by quantum curvature, the essential light cone is still well defined. In addition, the metric of space—its order of length, as it were—still has meaning. In fact, this metric turns out to be a proper solution to Einstein's field equations!

Is it possible to give a geometrical picture of how this complex space-time is changed by a single quantum of curvature? Remember that our twistor function of homogeneity +2 defines one half of the gravitational field, corresponding to gravitons of helicity −2. It turns out that this particular solution is curved in a left-handed sense but flat in a right-handed sense. A single graviton therefore produces curvature in the complex space-time *in one sense only*. Penrose calls this, by analogy with cricket, a "leg-break graviton." (A leg break is a ball that appears to be going straight for the bat but swerves away after hitting the ground away from the batsman.)

The twistor approach also makes it possible to create a contour integral around a twistor function with homogeneity −6; this will create a graviton with helicity +2. It turns out that such gravitons live in a complex space-time that is flat in a left-handed sense, but curved right-handedly! To complete this cricketing terminology, Penrose calls the task of creating such flat, left-handed spaces the "googly problem." (A googly ball is much more difficult to bowl, and after hitting the ground, the ball breaks towards the batsman or woman.)

So a single graviton can indeed be described both in the twistor and the space-time pictures, and in a way that displays the basic handedness or chirality of nature. Penrose's approach is particularly elegant, since it is purely gravitational and does not involve differential equations or algebra—it is another aspect of the "great dream." But this is only the beginning. The next step is to create a picture of a space-time in which both googly and leg-break gravitons exist. Remember that quantum theory allows for linear combination of solutions. Therefore the most general quantum graviton must be a linear combination of right- and left-hand helicities—googlies and leg breaks—complex spaces that seem to contain both combinations of curvature and flatness at one and the same time!

And finally, when very many of these quantum gravitons are admitted into the picture, it should be possible to generate a picture of a real space-time that is fully curved by the action of many gravitons. In the limit of large numbers of gravitons, we should get back to Einstein's classical description of a space-time curved by the action of matter and energy.

You may find this above description to be a curious combination of quantum and gravitational ideas. But this is exactly what Penrose was after. He did not believe that a conventional real space-time could be quantized in any simple way. Rather, gravity must change the whole structure not only of space-time but of the quantum theory as well. This must be contrasted with the superstring approach in which all of the beautiful and complicated new approach is essentially based upon the assumption that quantum theory remains unchanged right down to immensely short distances and even when the background space was indissolubly linked to the strings themselves. For Roger Penrose, however, gravity and quantum theory must transform each other.

The Ward Construction

General functions in twistor space have been used to create wave functions for the photon, neutrino, and graviton. An obvious extension would be to describe the important gauge fields that are used to explain the forces that operate between the elementary particles. R. S. Ward, who is also connected with the twistor group, showed that, on its own, twistor space is not general enough to explain the nature of these gauge fields. However, it is possible to give it an additional structure. At each point in twistor space, Ward therefore erected an additional geometrical structure, called a fiber bundle.

A fiber bundle can be pictured in the following way: We can learn something about today's weather by placing a thermometer in the garden. This registers the temperature, which is given by a single number. But suppose that the temperature outside varied from point to point—we would have to use a garden filled with thermometers—in other words, a space filled with numbers, each one giving the temperature. Physicists call this collection of numbers that give the value of, say, a temperature from point to point, a scalar field. A scalar field is like a contour map in which each point on the map corresponds to a number that is the height of the terrain above sea level.

Suppose, instead of measuring temperature, we want to measure wind speed and direction. We now erect a weather vane in the garden. This gives the direction and, with a bit of modification, the strength of the wind. In mathematical terms, it is as if an arrow has to be erected at each point in space. It is as if we stuck a pin in every point of a map's surface and on the top of this pin we mounted an arrow, something like a compass needle. In each particular case, these arrows point in the direction of the wind. Clearly the space has been enriched by the addition of all these arrows, or vectors. Physicists call this a *vector space*, and its structure is much more versatile than that of a scalar space.

Imagine now that the pins stuck in each point of the map carry on top of them a whole series of different arrows. It is possible to create a space, each point of which refers to a rich geometrical structure. This image gives something of the flavor of the fiber bundle that Ward added to each point in twistor space.

We saw earlier on in this book how the forces between elementary particles are associated with gauge fields—in other words, with gauge transformations in space-time. But now Ward was able to show that the gauge transformations in space-time have the effect, in twistor space, of causing the fibers, or configurations of tiny arrows, to become mixed up. But this mixing does not happen haphazardly, for the gauge transformations preserve a certain ordering among the fibers. Indeed, the pattern of the fibers at each point in twistor space is encoded in an ordered way by the transformations. In other words, the encoding of the fibers in twistor space corresponds to the gauge field transformations in space-time. In fact, we could say that this basic order or code within twistor space *is* the gauge field in space-time.

Ward's ingenious construction gives yet another way of understanding how nature's fields are created out of the complex structure of twistor space. For example, since the electromagnetic field can also be treated as a gauge field, it can now be thought of as a particular code in twistor space that is preserved by twistor transformations.

Again and again the twistor picture returns to Clifford's great dream that physics can be interpreted as geometry. Seeing massless fields in terms of contour integrals or encodings of structures in twistor space really comes down to the same thing: that nature's fields can be thought of in terms of the geometrical or cohomological properties of twistor space.

The Collapse of the Wave Function

Penrose's twistor approach to the problem of gravity has many implications. Edward Witten has written, "One of the striking developments in mathematical physics in recent years has been the twistor transform of the . . . Einstein and Yang-Mills equations." We have seen how these twistor transformations, induced by quantum theory, leave certain *local* structures in twistor space unchanged, and this means that the way *global* structures in space-time are encoded is unaffected. Points in space-time have secondary importance within the twistor picture. Indeed, it appears that the effect of quantization is to smear out these points and make them ambiguous. However, certain global structures such as the light cone appear to be unaffected. Recently Penrose has been thinking about how this global structure of space-time ties in with the quantum theory. Certainly it connects nicely to the idea of nonlocality in physics. In particular it makes a connection with those curious quasicrystals of two chapters back.

Remember that Penrose had shown it was theoretically possible to create a quasicrystal with a five-dimensional symmetry. In practice they can only grow when individual atoms have some sense of the overall form of the quasi crystal. Paradoxically this crystal cannot continue to grow unless it has a sense of its global form. The fact that global space-time structures can be encoded quantum mechanically in twistor space offers a hint as to how this could be done and shows how such a crystal could grow.

There is another way in which the nonlinear graviton has an important implication, and this is in the meaning of a quantum measurement. Quantum theory is essentially linear, which means that any superposition of solutions is allowed. Given two different solutions to Schrödinger's wave equation, any linear combination of these solutions is also valid. This gives rise to considerable problems when it comes to interpreting the results of a quantum measurement. Quantum theory purports to offer a complete description at the subatomic level of things, yet it does not tell us which one of a linear combination of solutions will describe the outcome of a particular quantum process.

This problem has been called "the collapse of the wave function." The wave function description contains a linear combination of many possibilities, while the actual result of a quantum measurement is a single realizable outcome. Somehow this linear combination collapses into a unique solution. The way this collapse occurs has puzzled physicists for over half a century. Some have gone so far as to suggest that it is the influence of the human observer or even consciousness. Others believe that all possible solutions are in fact realized, but in parallel universes. Penrose's approach is more elegant and less artificial, although whether it will satisfy the general physics community remains to be seen.*

A quantum measurement means that something that begins at the subatomic level must eventually be registered at our scale of things. It could be, for example, the click on a Geiger counter or the track of a particle left in a bubble chamber or on a photographic plate. But this registration clearly involves a magnification from the microscopic to the large scale. At some point in this magnification, Penrose suggests, the particular arrangement of microscopic bodies is sufficient to trigger the appearance of a single quantum of gravity. This graviton now makes a change to the structure of space-time; it induces a new global ordering. But this new order is nonlinear. It breaks the superposition principle whereby linear combinations of wave functions were earlier permitted.

The appearance of a single graviton is enough to collapse the wave function. Until the single graviton makes its appearance, many different configurations are actually

*The whole topic of quantum measurement and nonlocality forms the topic of the author's next book.

co-present at the microscopic level. But once these configurations have become magnified to a certain critical point, they produce a graviton, which then acts irreversibly to transform the whole space-time and, in this way, collapse the wave function. From this point onward, only a single outcome is physically possible.

It is possible to calculate how much mass is associated with the appearance of a single quantum of space-time curvature, which is (very roughly) around 10^{-7}g. This means that when a quantum process induces some rearrangement of matter or energy at this small scale—say a droplet of water measuring somewhere between a tenth and a hundredth of a millimeter—then this will produce enough curvature in space-time to collapse the wave function and produce a definite result. In this way the global ordering involved in a single quantum of gravitational curvature may be the key to resolving the quantum measurement problem.

The measurement problem has been around for fifty years now, and Penrose's new approach has yet to be debated by the general physics community, so it is difficult to evaluate its significance. In fact, Penrose himself would rather not call it a solution to the measurement problem, more a "pious hope." Nevertheless it does expose a basic problem in quantum theory: Can quantum theory really be interpreted consistently without also including the properties of space-time? Is quantum theory an entirely separate and self-consistent theory that stands apart from general relativity, or should the two theories be viewed as the respective limits of some much deeper and as yet undiscovered theory?

More recently Penrose has even begun to think about this *collapse of the wave function* in terms of a flipping or jumping between twistors. This is not so much a new theory of quantum measurement as an "idea for an idea." Essentially Penrose asks what happens if the twistor space itself is curved. A quantum system may quite happily be represented by a particular twistor, yet because of kinks or bifurcations in the underlying space, the system may jump from one twistor to the next. The overall effect appears to be a sort of jump within the wave function.

This is the sort of idea that may or may not work out—if it does, it would seem to suggest that even the collapse of the wave function has a geometrical interpretation. Moreover it would indicate that the structure of quantum theory is affected by the structure of twistor space and therefore, indirectly, by the structure of space-time. Again Penrose has brought us to the point where quantum theory and space-time interconnect and become inseparable. His overarching vision is that neither theory is really fundamental but that both must emerge out of some deeper, and yet to be discovered, form.

Conclusion

With the help of his twistor formulation, Roger Penrose has been able to extend part of the great dream. His geometry founded on extended objects is able to make significant connections both to quantum theory and to relativity. It is able to describe massless fields in a new, topological fashion, and to give a meaning to quantum curvature at the quantum level. In the long term, it may provide a picture of the elementary particles and their respective masses and symmetries. It may also shed light on the overall structure of space-time and its origins in the big bang.

In a 1986 edition of the British scientific journal *Nature,* George Sparling showed how five of the important theories of physics could be written down in simple terms using the twistor approach. These included Einstein's classical formulation of gravity, supergravity theories in four and eleven dimensions, and combinations of Einstein's gravity and Maxwell's electromagnetism. In each case the information about these fields appears as a coding of fiber bundles in twistor space. In the opinion of Lance Hughston, this geometrical principle may well turn out to be as important as the principle of least action that was discussed earlier in this book. Twistors, it appears, may well have a profound impact upon all of physics.

Twistors therefore provide one road toward a truly unified theory of physics. The other is given by heterotic superstrings and their intimate connection to the elementary particles, gauge fields, and background space-time. In the final chapter, we shall explore some ways in which these two approaches may meet up. In addition we shall discover something about the frontier research that is being carried out in superstring theory.

10
Into Deep Waters

Postmodern Physics

PHYSICS BEGAN IN this century with Ernest Rutherford firing alpha particles at a piece of gold foil, and now it appears to be ending the century with groups of theoreticians gathered round a blackboard. It is as if, within a couple of generations, physics has undergone a profound and irreversible change. Rutherford was primarily an experimentalist, yet he also created new scientific models of the atom. Over the next two or three decades of this century, theory and experiment continued hand in hand as they had done for the past 300 years. Sometimes theory lagged behind experiment, with new observations suggesting modifications and calling for fresh intuitions. On occasion, as with the general theory of relativity, which predicted the bending of starlight as it passes close to the sun, a new theory could stimulate experimental observations. But, generally speaking, there was no great distance between what could be measured and observed and what could be thought about and described mathematically.

Today things are very different, for the advanced theories of physics have little direct connection with anything that can be measured, and those experiments which are suggested by the theory are probably decades away from being designed. Theories today are really emerging out of other theories, and their testing ground is no longer the experimentalist's laboratory but aesthetics, mathematical consistency, and their interrelationship to yet other theories. In addition, the mathematical language in which these theories are expressed has become so advanced that it is no longer always possible to give simple visual illustrations of what the theory means. Indeed, that has made this book difficult to write: in going from chapter to chapter the concepts and ideas have become more and more abstract, so that at times I felt as if I were trying to explain Bach's *Art of the Fugue* in words.

Yoichiro Nambu, the creator of the original string theory, has called this situation "Postmodern Physics." Theory has moved so far ahead of experiment that, he suggests, physics must now develop in new ways. When a new theory is created, rather than thinking in terms of crucial experiments and observations, physicists have to begin by investigating the theory's formal mathematical structure. The theory and its mathematical language are probed, recast, and related to other theories. Eventually it will be possible to discover its most fundamental form. So, rather than seeking immediate contact with experiment, the idea is to advance the theory, expand its scope, make it consistent, and relate it to other theories. Eventually it may turn out that this whole process leads to predictions or to observations. But such observations need not be direct. Crucial tests may be first established in terms of other theories.

For example, the predictions being made by superstring theory involve energies that are so remote from anything that we can produce in the laboratory as to be beyond all possibility of testing within the next few decades. Nevertheless it is possible that the cosmological consequences of the theory can be tested for their consistency with other, cosmological, theories of the universe, and these in turn may ultimately make contact with observation.

If Nambu is correct, then physics has set itself on a new course in which its theories will become increasingly abstract and remote from direct observation. In addition it may be as difficult to give simplified accounts of these new theories as it would be to explain the frontiers of abstract mathematics. So much for us popularizers of contemporary physics!

How has Nambu's idea worked in practice? It seems to apply quite well in the case of superstrings. Remember that Green and Schwarz's great breakthrough was partly triggered not by some new experimental observation but by the ten-dimensional point particle theory advanced by Alvarez-Gaumé and Witten. Their immediate crucial test was internal consistency, freedom from infinities and anomalies, logical inevitability of the arguments, and mathematical aesthetics. The test they now face involves things like discovering the exact mechanism for compactification and symmetry breaking. Rather than the theory being measured against some novel experiment, it is required that it should reduce, at atomic distances, to certain features of successful point particle theories. Indeed, there is little possibility of superstring theories making direct contact with experiment within our lifetimes. The most crucial test may lie in their cosmological predictions, for example, how superstrings modify the theory of the big bang origin of the universe.

Nambu's description of postmodern physics seems substantially correct. Physicists today are concerned with investigating superstring theory, recasting it, probing, making connections with other approaches like twistors, attempting to refine the mathematics, exploring theoretical implications. Indeed, these frontier topics in superstrings form the contents of the present chapter and will show something of the deep waters into which theoretical physics is now heading.

But before we take the plunge, I would like to offer my own observations on Nambu's postmodern physics. It seems to me as if physics, as a social dialogue about reality, has moved closer to the flexibility and properties of language itself. It is only in the baby-talk stage that we seek to make an immediate relationship between words and objects. An ostensive or "pointing" role of language implies that everything has a name and that language points directly to objects or facts in the external world. This may be the sort of thing a person becomes involved in during the early stages of learning a new language, but soon the speaker enters a far richer world in which language has many more uses.

Language is essentially a means of communication, not only socially but also internally as we reflect on our ideas and carry out private, inner dialogues. Its syntax and semantics are the tools with which we can perform a great variety of tasks. The meaning of what we say does not generally correspond in any direct way to objects or facts in the outside world; rather, meaning lies within language itself, that is, within the whole activity of communication within our society. Language is concerned with meaning, and meaning is generated out of social and internal use of language.

This is not to say that an external reality has no significance within language. We are totally free to say just what we like; nevertheless, if we want our actions to cohere within society and to make sense in the physical world, then at some level the way we use language has to correspond to the way individuals, societies, and nature operate. Without getting into a major debate about the meaning of the world *reality*, let us say that we are using it here in the sense of the way we act and communicate individually and socially. Clearly this must go beyond the purely subjective, for if what we all agree upon does not cohere with the way the physical world works, then at some point we are going to get into trouble. Reality, therefore, implies an ongoing and ever-changing two-way dialogue that involves the individual, society, and nature—and science is an essential part of this dialogue.

While our linguistic communication normally ranges

far beyond the purely representational, it must at some point stand in accord with the physical nature of the world. Nevertheless within this context we have great freedom to create new meanings and new dialogues, and the very ways we come into contact with nature and interpret our observations are profoundly conditioned by what we think and say. I would suggest that this is exactly the status of modern physics. It has freed itself from having a purely representational role and is now much closer to a language than to making simple pictures of an objective, external reality.

Modern physics is about theories, which are a form of social communication for discussing reality. Its currency is mathematics and mathematical models. Its main concern is for coherence and internal consistency, for freedom of expression, and for aesthetic value. Theories are increasingly about other theories, just as language is about language—in the sense that the structure and meaning of language can only be discussed using language itself. The meaning of theories does not so much lie externally in terms of data and observations as internally, within the whole complex network of modern theoretical ideas.

This is not to say that physics is free to do anything it likes. For at some point this whole theoretical dialogue must make connections with the external world, through the experimentalist's laboratory. Yet even the way we perceive the world and design our experiments is, on the social level, determined to a very great extent by the theories themselves. Nevertheless it is possible at some point for the whole structure to come tumbling down and for the theoretical world view to fail to cohere with experience. In the early days of science, this was a fairly direct process. Theories were relatively simple pictures of reality and easily subject to a crucial test. But today experimental inferences are becoming more and more indirect, since each piece of data needs to be theoretically predigested.

So while there is still a meaning to objectivity in postmodern physics, and a theory can stand or fall on the basis of experimental confrontation, I would argue that the particular beautiful network of theories we create need not necessarily be unique. It could be that there is some quite different way of proceeding, just as there are many ways in which a painter can portray the same scene or character, each one involving some essential truth and consistency about the subject. It may well turn out in the next century that physicists can present a number of alternative theoretical accounts for the same network of ideas, concepts, and observations. In turn, deeper interconnections between these different theories will be exposed so that the whole abstract structure will change yet again. Clearly a physics of this nature can have no end but will be involved in a ceaseless process of transformation, clarification, and unfolding. Its goals may well turn out to be aesthetic rather than practical.

So physics moves closer to language in its flexibility for dialogue. If this is true, then the frontiers we discuss in these chapters are indeed closer to Bach's *Art of the Fugue* than to, say, Newton's theory of color. Twistor theory, superstrings, and Bach's meditations on the nature of fugue are all artworks, whose meaning lies partly in their beauty and in the way in which they open up new levels of communication and exploration and, in the scientific case, in the way they cohere with our experience of nature.

Postmodern Superstrings

Now that the initial excitement about superstrings is beginning to die down, physicists are taking a long, hard, critical look at the whole theory. This involves analyzing the whole structure and implications of the theory and attempting to put it on a much firmer theoretical foundation. The current program includes trying to free the theory from its attachment to perturbation theory and a flat background space-time, applying certain insights gained from the twistor approach, discovering ways of exploiting the relativistic covariance that is inherent in the theory, developing a proper string theory, and creating a rigorous account of how its initial ten-dimensional space-time is compactified so that the various symmetries of nature and elementary particle masses can emerge. We shall look at some of these frontier topics in this final chapter.

Twistors and Superstrings

Edward Witten has suggested that twistors may well be "the proper starting point for understanding the geometrical meaning of superstring theory." If this turns out to be true, then it means that a profound connection could be made between two of the most exciting theories in modern physics. By combining their respective strengths and insights, it may be possible to advance our understanding of, for example, the elementary particles in purely geometrical terms.

Lane Hughston and W. T. Shaw have been exploring the connections between strings and twistors. Remember the duality between the twistor space and space-time pictures? Points in certain regions of twistor space become lines in space-time; lines in twistor space become points in twistor space. Now Hughston and Shaw have extended this picture by showing that a very general curve, called a holomorphic curve (or, in some cases, two such curves), in twistor space corresponds to a relativistic string in space-time. In other words, the most general of the well-behaved structures in twistor space can be used to generate the relativistic starting point for a superstring.

In essence, the holomorphic curve in twistor space corresponds to a *minimal surface* in space-time. Recall that this minimal surface is swept out by a string as it moves through time. In writing down the equations for a string, it was necessary to satisfy an action principle—

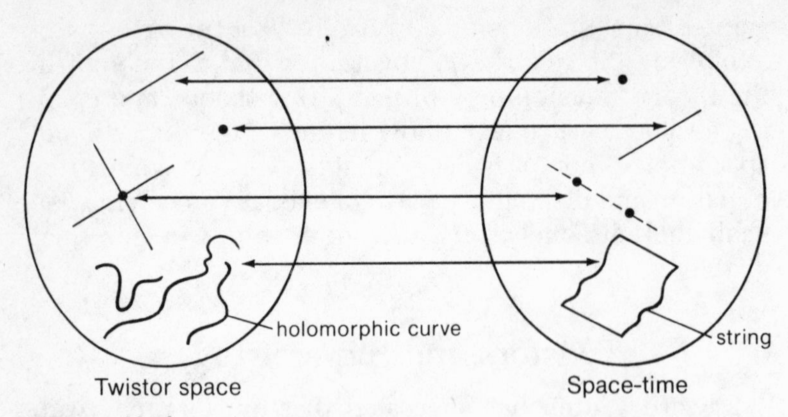

Figure 10–1

the principle of least action—which, in the relativistic picture, produces a minimal surface. Not only do fundamental structures in twistor space provide the starting point for the string, but they also have a profound connection with the action principle of physics.

Let us follow the logic of the latter argument. Twistor transforms, such as those brought about by quantum curvature, will generally destroy the global properties of twistor space so that straight lines are broken apart. But these do leave the structures of general holomorphic curves unchanged. In fact, one could say that holomorphic curves *are* the manifestations of mathematical good behavior in twistor space. But in the space-time picture, these curves also correspond to minimal surfaces, and these are the histories of structures that obey the principle of least action in their dynamic unfolding. In other words, good behavior in twistor space is equivalent to an action principle in space-time, and this means the basic laws of physics (for the laws of physics can be reexpressed as action principles). This equivalence suggests that it may be possible to recast physics, including superstring theory, in such a way that its starting point lies in the geometrical structures of twistor space.

Hughston and Shaw have shown that this correspondence can be created not only in a four-dimensional space-time but in several other dimensions as well, including the ten dimensions favored by superstring theorists. In each case it is possible to generalize the idea of a minimal surface in space-time, based on the underlying holomorphic structure in twistor space. The next step will be to quantize the string. Already Hughston and Shaw have attempted this quantization procedure in a four-dimensional space-time.

There is yet another spin-off from Hughston and Shaw's work, and it concerns what are known as cosmic strings. During the very early stages of formation of the universe, different regions formed that were separated from each other by boundaries called cosmic strings. The effect is a little like growing a dense set of crystals; where the crystals touch, each with a different orientation, growth stops and a boundary forms. In the cosmic case, it looks as if the twistor holomorphic curve provides a good space-time description of these cosmic strings. Since these strings have recently become such a hot topic, they will be discussed in a section of their own a little later in this chapter.

If the properties of twistor space can be used to generate the (prequantized) strings themselves, is it possible to draw upon yet other aspects of Penrose's twistor theory? Edward Witten is impressed by the way in which the twistor picture sheds new light on gravity and the gauge fields of nature. Rather than thinking about them as fields in space-time, Penrose was able to deal with much deeper structures in twistor space. Could this approach be extended to embrace superstrings as well?

The foundation for such an approach is a supersymmetric twistor called a supertwistor. Since Witten feels that a ten-dimensional space-time is the natural arena for supergravity and the gauge fields of nature, the next step is to extend the twistor formulation so that it can generate a complementary picture, not in our four-dimensional space-time but in the ten dimensions of superstring theory. In this way it should be possible to exploit all the power of twistor transforms in twistor space and express this in terms of superstrings in a ten-dimensional space-time. Here Witten parts company from Penrose, who has aesthetic and physical arguments for remaining with a four-dimensional space-time. (However, Witten's views may have recently changed, since some very powerful mathematical arguments have been demonstrated to work only in four dimensions. Somehow the properties of a four-dimensional space are very special, and this may be connected to the reason why the space-time in which we live happens to be four-dimensional.)

It has turned out that things are more difficult in the ten-dimensional approach. Nevertheless Witten was able to generalize the twistor picture to describe gauge fields and even to make the first steps toward a gravitational theory. However, at this stage it did not prove as easy to move back and forth between the complementary space-time and twistor space pictures as was done in the four-dimensional theory.

Witten's idea is to try to go beyond the conventional flat space-time formulation of superstring theory and understand what it means for a massless point particle or a string to move in a curved rather than a flat space-time. The space-time histories of string in a curved space-time are given by the twistors. But now the curvature itself introduces general deformations into the picture. For the dynamics of the string to continue to make sense, it is necessary, Witten argues, to retain some sort of structure that refers to locally flat regions. This practice of being able to cover a curved space with a series of overlapping flat spaces has been met before; in general, however, these flat spaces will not join up smoothly at their edges. Even though there are general deformations in twistor space, it is still necessary to keep some information about local conditions.

But Witten's approach poses difficulties, and it looks

as if some additional structure must be added to the twistor picture. At present it is not clear how this can be done, and fresh insights may be needed to press ahead with this approach.

Yet another connection between superstrings and twistors, albeit at the purely mathematical level, stems from the work of Singer, Penrose, and Hodges. It was Michael Singer who suggested that the trouser diagrams that are so essential to the whole superstring approach could have a direct correspondence to twistor diagrams in twistor space. Starting not with a closed loop in superstring space but with a piece of twistor space, the special region *PN*, it becomes possible to have these twistor space regions interact by means of the various trouser diagrams. By borrowing the various powerful techniques that have been used for superstring diagrams, it may be possible to make a breakthrough in understanding the meaning of twistor diagrams.

Compactification

Consider superstrings in retrospect. The theory grew out of an attempt to create a unified picture of nature. Its basic philosophy is that the universe emerged, through a process of progressive symmetry breaking, from an initial state that had a very high degree of symmetry. Historically this approach evolved from the first quark models of the hadron, with their SU(3) symmetry, the unification of electromagnetism with the weak nuclear force to create a single electroweak force, and finally the grand unification of electroweak and gluon forces. It was assumed that all three forces of nature were unified at the moment of creation of the universe and represented by the symmetry SU(3) × SU(2) × U(1) but that within an incredibly short time this grand symmetry became broken and the various elementary particles differentiated, each acquiring a different mass. To this grand unified model was added supersymmetry, which links the fermions and bosons and opens the possibility for a further form of unification, this time involving the gravitational force in the form of supergravity.

But this grand picture must also reproduce the chirality or basic handedness of nature. It was in trying to keep all these various balls of symmetry and chirality in the air at the same time that conventional point particle theories met their downfall. While the demands of grand unification and supersymmetry had forced theoretical physicists to leave our familiar four-dimensional space-time for higher-dimensional formulations, it still did not seem possible to satisfy the requirements of symmetry and chirality and at the same time end up with a theory that was free from anomalies and infinities.

It was at this point in 1984 when theoreticians were thoroughly discontented with their conventional point particle approaches, that Green and Schwarz came up with their superstring theory. Formulated in ten dimensions, the theory was not only chiral, supersymmetric, and included the grand unified symmetry, but was also free from all those inconsistencies, anomalies, and infinities that had plagued earlier theories. Green and Schwarz's theory was formulated using the symmetry SO(32) and featured both open and closed strings. The various forces of nature were represented in terms of the splitting and joining of strings.

Shortly afterward came Gross's heterotic theory of closed strings in which gravity is pictured through the interaction of closed loops with their background. The other forces of nature are also supposed to emerge when the ten-dimensional space of the heterotic strings compactifies down to four.

Ten-dimensional superstring theory has become the triumph of modern theoretical physics, for it is able to satisfy so many different constraints and requirements in such an economical way. But to make contact with our own world, its results must be somehow boiled down to our own space-time of four dimensions. This is where the whole issue of compactification comes in, for it is by discovering just how to curl up or contract six out of the initial ten dimensions, and in exactly the right way, that superstrings must stand or fall. The question is, starting with a space-time of ten flat dimensions, how is it that we experience only four dimensions? Or, conversely, we could ask why the universe was initially created in ten incredibly small dimensions yet only four of these dimensions expanded and flattened out, leaving the other six in their highly compacted form.

Not only do we have to explain the mechanism of compactification in order to express the theory in our own four-dimensional space-time, but in addition we have to explain how the initial heterotic symmetry of E8 × E8 became broken to the point where the familiar symmetries of elementary particle physics begin to manifest themselves. The process of compactification must show how the four forces of nature emerged out of a single unified picture, each with its own strengths, and why the elementary particles have their own particular masses.

But even this is not enough, for string theoreticians must also prove that this process of compactification is stable, so that the hidden dimensions do not suddenly decide to expand explosively, or to oscillate, or that some of our own four flat dimensions do not themselves decide to compactify! For example, if the tiny radius of the compactified dimensions were to begin to fluctuate, then the forces of nature would change strength, elementary particle masses would alter, and the whole of subatomic physics would be thrown into a state of chaos.

According to quantum theory, we expect such quantum fluctuations in the radius of the compactified dimensions. In addition, general relativity suggests that the curvature of the compactified space should not be stable. Yet all the forces and masses of the elementary particle world are so finely tuned that even the smallest change in compactification would throw the whole order of the subatomic world into disarray. If this happened, it is unlikely that any large-scale structure such as molecules, rocks, life forms, planets, or stars could survive. But we know that life, the earth, and indeed the whole solar system have been in existence for a very long time, so we can be certain that these compactified dimensions must have remained stable, at least since the first few fractions of a second after the big bang creation of the universe. One theoretical problem is how to explain this stability.

The major step in unfolding superstring theory is to explain how this very elegant ten-dimensional theory of extended objects ends up, at elementary particle distances, looking like a point particle theory with four forces and elementary particles having a particular pattern of masses. In other words, the fine details of compactification have to be worked out. It may turn out that such questions cannot be adequately answered while the theory is in its present form and that what is required is some deeper understanding, for example, a quantum string theory that is freed from the perturbation theory approach and founded on some more fundamental principle. At present, however, the topic of compactification is attracting some deep mathematical analysis, only a sketch of which can be given here.

Orbifolds

At present Green and Schwarz's and Gross's superstring theories are formulated in a flat ten-dimensional space. Essentially this is a ten-dimensional Minkowski space-time and could be denoted as M^{10}. But the theory must end up describing particles and fields in a four-dimensional space-time M^4. (At some point the whole theory may be generalized into a general curved space-time, which should emerge naturally out of the theory itself.) So this means that somehow M^{10} must compactify or break up into M^4 and K, where K is some as yet unknown six-dimensional compactified space. The size of the dimensions of K must be so incredibly small that they cannot be seen, even at the elementary particle level. The assumption is that the radius of the dimensions in K is the same as the superstrings themselves.

To explain the curled-up dimensions of this compactified space, K, the image is sometimes given of a piece of macaroni. Macaroni is a cylinder whose dimension of length is visible but whose thickness or radius cannot be seen when viewed from a distance. Is it possible to do something similar in the space K and end up with a generalized picture of "macaroni dimensions"?

Let's start with a chessboard and begin to curl up its dimensions. First we join the top of the board to the bottom and end up with a cylinder—a length of macaroni—in which we make the radius very small indeed. We have succeeded in compactifying one of the dimensions. Now join the left-hand side of the board, or cylinder, to the right. The result of joining the ends of the macaroni is to create a torus, T, or doughnut. If this second radius is also very small, then we have managed to compactify the two dimensions.

Now generalize this procedure from a torus T^2—which

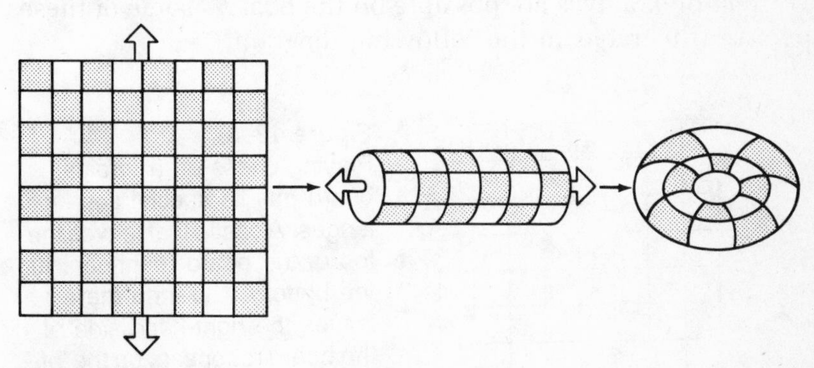

Figure 10–2
A chessboard can be rolled up into a torus.

is a two-dimensional surface in three dimensions—to a torus T^6 which involves six compactified dimensions. In this way the original flat ten-dimensional space-time M^{10} is broken down into four flat dimensions and a six-dimensional torus, each dimension of which has an incredibly small radius. M^{10} is compactified to $M^4 \times T^6$.

This is the simplest way in which we can compactify or curl up a ten-dimensional space and recover the four dimensions of our space-time. The six dimensions of the torus are so tiny that they can never be seen. The only problem is that this process is too simple. Curling up space like a doughnut does in fact break the initial symmetries of the ten-dimensional superstring theory, but not in the right way, so the spectrum of particles and the gauge forces it creates have no connection with reality.

What is needed is a richer way of folding up space than the simple chessboard picture. Some people will actually play torus-chess on a flat board. Imagine moving your castle to the top of the board. Since the rule for making a torus is to join the top and bottom of the board, this means that your castle on leaving the top of the board will suddenly reappear at the bottom. Since the two sides are joined as well, this means that even more complex moves are possible on the board—some of these are illustrated in the following diagram.

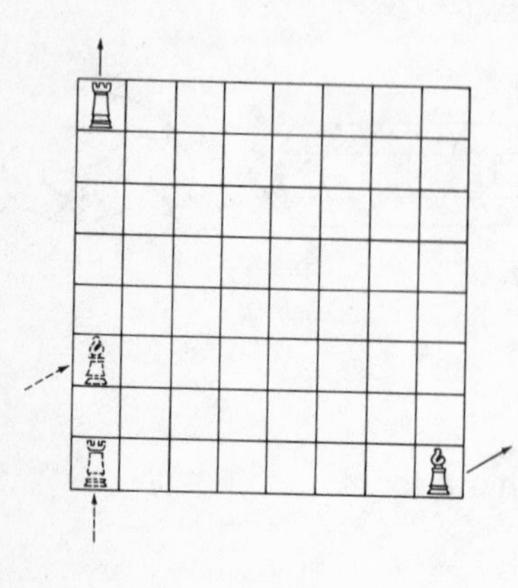

Figure 10–3
Playing chess on a toroidal board has its special challenges. A castle that leaves the top of the board reappears at the bottom. A bishop that leaves the right-hand side of the board reappears on the left-hand side.

This idea can be extended to an orbifold space in which, in addition to the torus properties, it is possible to have an additional connection near an edge. See how the piece leaves the center of one edge and immediately reappears again. Clearly the point X has a special role to play. In fact, with three special points on the board, the compactified space looks like this:

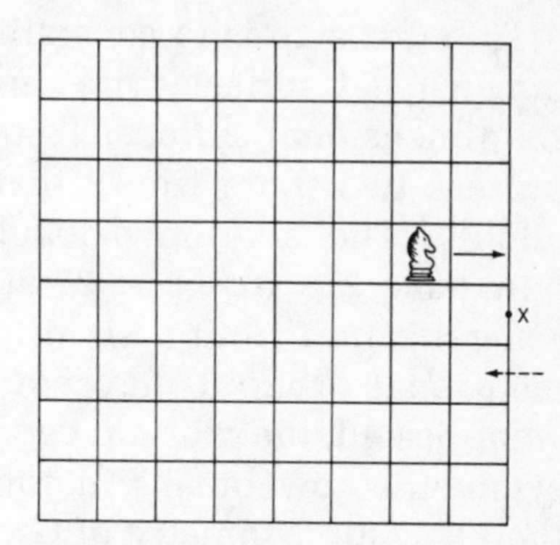

Figure 10–4
This chessboard has a special singular point along one of its edges. A knight leaving the board just above this point reappears just below. Such singular points are used in the construction of an orbifold space.

The special points are called conical singularities—singularities because the properties of the chessboard, or orbifold, blow up at these points. The result is a curious cross between a torus and a sort of three-dimensional triangle that mathematicians call an orbifold.

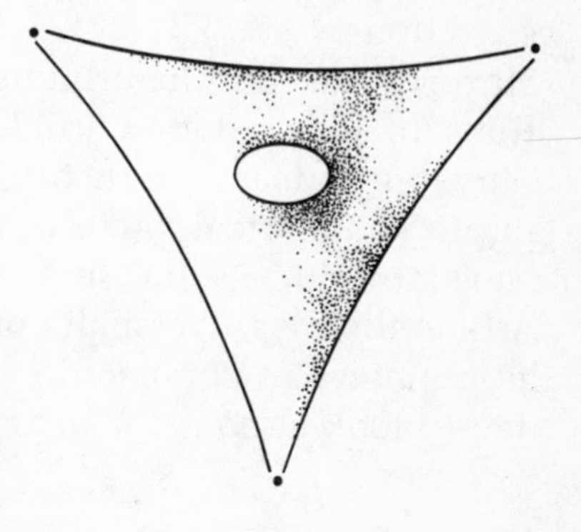

Figure 10–5
This orbifold space consists of a torus with three special, singular points. Like the torus, it is a two-dimensional surface in a three-dimensional space. By combining three such orbifolds together, it is possible to generate a six-dimensional space with $3 \times 3 \times 3 = 27$ singular points. Such orbifolds are used in string theory to represent the space generated when the original ten-dimensional space compactifies.

The interesting thing about this orbifold is its three conical singularities. In a torus, properties change very smoothly as we go from point to point, and all points on its surface are equivalent. The same thing holds on an orbifold, except for its three conical singularities, and this additional new structure will give the orbifold the new properties we want.

The next step is to generalize our orbifold from a two-dimensional surface (in a three-dimensional space) to a six-dimensional surface. To do this we repeat the whole process two more times; that is, we combine three orbifolds. The six-dimensional compactified space will now have $3 \times 3 \times 3 = 27$ special points. Although the generalization from a six-dimensional torus to a six-dimensional orbifold may not seem all that much more complicated, the orbifold certainly seems to work. Supersymmetry now breaks in the correct way. In addition, the E8 × E8 symmetry of Gross's heterotic string theory begins to break on the orbifold in just the right way. That is, one of the E8 groups breaks to SU(3) and E6. Of course, the E6 itself has yet to be broken into an even finer structure, but already we can see that, as a result of folding up part of the superstring space like an orbifold, the theory will produce quarks and leptons in several generations.

It turns out that the orbifold predicts thirty-two generations of elementary particles. This is clearly far too many, but at least the theory is on the right track, and a simple geometrical picture of curling up a space will indeed produce some of the features of known point particle theories. We shall come back to this question of the geometry of compactified spaces later, but now it is time to look at some of their implications.

The Problem of Mass

Superstrings live in a ten-dimensional space-time that contains various coordinates that describe the bosons and fermions, chirality, and an overall symmetry. Within this ten-dimensional space, it is possible to write down equations that resemble those of conventional fermion elementary particles in the more familiar four-dimensional theory. When compactification occurs, not only does the E8 × E8 symmetry break, but also the "elementary particle" equations split into a four-dimensional and a six-dimensional part.

At this point something very interesting happens to the "elementary particle equation," since from the perspective of our everyday four-dimensional space-time, its two parts look familiar. The part written in four dimensions looks just like the quantum equation for an electron or proton, while the part written in six dimensions looks exactly like its mass. In other words, the solutions to this equation in six dimensions, its pattern of notes as it were, turn out to be the actual masses of elementary particles! While the electron, proton, and neutron appear to live out their lives in our own four-dimensional space-time, their actual masses are an expression of their additional hidden existence as superstrings in the six compactified dimensions.

But how do these masses compare with those found experimentally? It turns out that the smaller the radius of the compactified space, the larger the spacing between the notes on the mass scale. When spaces are measured in terms of the lengths of a superstring, the spacing is enormous—masses correspond to energies of 10^{13} GeV and are vastly beyond anything that could ever be seen in elementary particle experiments. Particles with such masses could only have existed within the first instants of the creation of the universe.

Superstrings are fluctuating, two-dimensional world-sheets in a ten-dimensional universe. After six of these dimensions compactify, the strings look, at the subatomic level, like elementary particles moving in a four-dimensional space-time but whose shadow, compactified nature shows up as mass. But since the size of this mass depends on the smallness of the compactified space, it turns out that all the mass "notes" are far too large to make any sense in our world. In other words, the only chance is for the elementary particles to have no mass at all, to correspond to the very first (zero) note of the mass scale!

At first sight this appears absurd. We know that all the elementary fermions—except for the neutrinos—have mass. On the other hand, it is certainly true that these masses *are* virtually zero when looked at from the perspective of energies comparable with grand unification. Modern physics tells us that the universe is created in a highly symmetric state but these symmetries are broken or hidden in our everyday world until we reach very high certain energies. The energies at which these symmetries exist, the grand unification energies, could be thought of as the natural energies of our universe. If super science fiction beings could live at such energies, they would see a highly symmetric universe, with only a single force and particles whose "notes" of mass were created out of the compactified space. To such a being, our own world would be virtually invisible, an incredibly tiny baroque decoration. From this superperspective, the masses of electrons and protons would indeed be virtually zero.

Since, from the perspective of this superstring scale, all excitations have enormous mass, the burning issue is how to explain why the protons and electrons of our experiment do not pick up masses of 10^{19} GeV, but to all intents stay massless until the very last moment!

With luck it may turn out that these initially massless electrons, protons, and neutrons will eventually pick up their own negligible (negligible, that is, in comparison to the enormous energies of the superstrings) masses in some other symmetry-breaking process. In other words, the real masses of the elementary particles are extremely fine adjustments to the basic superstring theory. All the physics we have ever studied, our own lives, and our consciousness are really enormously fine corrections to an underlying level.

In fact, this picture is in accord with what we have already learned about the ways in which nature works. Our consciousness involves fine-tuned chemical and electrical processes at energies that are negligible when compared to the more violent processes that can occur when molecules are created. In turn these molecular processes are negligible when viewed on an atomic scale. Atoms themselves are fine structures when compared to the elementary particles. And now we have discovered that the elementary particles themselves are negligible corrections to the underlying superstring picture. Nature, therefore, involves a whole series of hierarchies, each one appearing as a tiny adjustment or correction to an underlying level. Some scientists have argued that new levels of complexity should be viewed as emergent structures, novel forms that were not expected on the basis of an underlying level. In this sense a reductionist picture is always limited, for it never prepares us for the complexity and richness of the next level of organization.

The universe we live in is a very fine correction, an almost negligible decoration upon the underlying superstring theory. Here an image may help. Think of a television set. It is plugged into the wall, and several amps of electricity at 110 volts enter the set. This current supplies the natural energy level of the television set. Yet superimposed on this energy is a virtually negligible energy—the tiny fluctuations in current that are picked up in the antenna from the electromagnetic signal broadcast by the television station. The energy that comes from the antenna is negligible when compared to that drawn into the set from the wall plug. One would be justified in ignoring it as being unimportant. Yet this negligible energy carries information and this information shapes the much greater energy of the television and gives it form. The result is a picture and the sound. The much greater energy of the television set itself, which has a simple and symmetrical order, is modified by the negligible energy of the signal, which has a complicated form or order.

In a similar way, the great energy of the ten-dimensional superstring theory is somehow shaped and formed by processes at a vanishingly small level. These processes are of such complexity and subtlety that they produce the rich physics of our universe: they create elementary particles; these particles form into atoms; atoms into molecules; molecules into solids, liquids, and gases, which in turn form stars, planets, trees, animals, and our brains.

We could also take the anthropic perspective, that the universe we know involves the fine details of this structure. We see the universe through very fine energy probes, light, x-rays, sound, etc., signals whose energy is negligible when compared to the vast energies of superstrings and grand unification. But this also means that we can only "see" those elementary particles and phenomena which come within our range. If there are indeed more massive particles around, then we simply cannot see them. "Our" universe exists at fine energy scales and at such complexity that conscious beings can evolve and make the sort of statements I am now making. Change the energy scale, break the symmetry of the ten-dimensional superstring space in a slightly different way, and this book is not written.

So superstring theory has to stop elementary particles from having enormous masses, and this means giving them masses that are initially zero. The next step is to add a very fine structure of masses and reproduce the known elementary particles. It turns out that to make sense of experiments, the quarks and leptons have masses that are no bigger than a few hundred GeV, far from the 10^{19} GeV of the compact space. That means we must take the first, zero-mass note in the compactified space and make sure that all the fermion particles continue to stay massless as symmetries are broken, until the electroweak interaction is broken into electromagnetism and weak interaction. If mass is picked up before that point, then it will be far too large and the universe that would result would have very little connection with anything we understand. Following the breaking of the electroweak symmetry the elementary particles can pick up their various masses. This picture of mass is not unlike that proposed by Penrose, in which the universe begins in a massless, conformally invariant way.

But this is a great puzzle—why is the scale of the elementary particle masses, a few hundred GeV, so enormously smaller than the natural scale of 10^{19} GeV? This is one of the great questions of modern physics. It tells us that the elementary particles can only begin to get their masses at around the point that the electroweak interaction breaks down and the intermediate vector bosons, W and Z particles, first appear with masses.

The smallness of the masses of the elementary particles is also connected with chirality. If there were no basic handedness in nature, if left and right were totally equivalent, then right- and left-handed electrons could interchange with each other and in this way pick up additional mass. Calculations show that such masses would be too large. Once again the elementary particles are trying to attain very big masses, and the only thing that stops them is that chirality prevents left- and right-handed electrons from mixing. Right until the point at which the electroweak interaction breaks down, these two sorts of electrons are kept apart.

It is thanks to chirality that we have an observable universe at all! If nature were more symmetrical, then the masses of all the elementary particles would be enormous. Again we have a deep explanation for chirality—if it didn't exist, then the universe as we know it would not be possible. Penrose gives an alternative explanation: the most basic way in which fields can be written down in a complex space has to be chiral. Possibly the two ways of seeing chirality will eventually interconnect.

Now back to mass and compactification. We want to prove that it is indeed possible to have massless electrons, protons, etc., right down to the very last moment. This means that the spectrum of "notes" in the compactified space must start at zero and not at some finite number. We can prove that this will happen if the dimension of the space is $4n + 2$, where n is some number. In the case of $n = 0$, we have two dimensions, the dimensionality of a superstring world sheet. The next possible dimension requires $n = 1$ and is six dimensions. So six dimensions seems to be the first dimensionality in which elementary particle masses make any sense at all.

This is exciting, for it gives a dimensionality to our world. It says that if the universe is created out of superstrings in ten dimensions—and this appears to be the only sensible dimensionality in such a theory—then such a space *must* compactify down to produce a four-dimensional space-time; otherwise there can be no elementary particles with negligible masses. There is a compelling reason as to why the space we live in has to be four-dimensional. One of our outstanding questions of Chapter 1 has been answered. Of course, we still have to fill in all the fine details and show how the final acts of symmetry breaking produce the known elementary particles along with their particular masses.

Up to now this discussion of mass has been restricted to fermion particles, like electrons, protons, neutrons, and neutrinos, which make up our material universe. But the force-carrying bosons also must be considered. These must emerge out of the superstring theory looking like gauge fields. Again it is necessary to study the whole compactification process and see if it will also work for these gauge fields so that they appear with the correct symmetry out of a combination of six curled-up dimensions and four flat ones. As with the fermion case, these are difficult, technical matters which are still being investigated.

In summary, therefore, compactification of six out of ten dimensions enables us to write down an equation containing two terms, one of which looks like a conventional elementary particle in four dimensions of spacetime, while the other, which is defined in the curled-up space, looks like the particle's mass. However, the spectrum masses produced by the compactified space are so incredibly large as to have nothing at all to do with what we traditionally call elementary particle physics.

The problem, therefore, is to begin with the lowest note of the spectrum and have only massless particles. These particles continue to remain massless right down to the energies at which the electroweak force breaks its unification. At this point the various elementary particle masses can make their appearance as very tiny corrections to their initial zero masses. Accompanying the appearance of mass will be one last stage of symmetry breaking into the final symmetries of the various families of elementary particles. Working out the fine details of all this is a current problem in superstring theory called *phenomenology*.

Calabi-Yau Spaces

Let us look at another way of thinking about compactification. At the start of this general section, we looked at two ways of compactifying a space. One of these produced a torus, whose structure was not rich enough; the other created a curious figure called an orbifold. The orbifold certainly had some of the correct properties; for example, it seemed to be along the right road for breaking some symmetries and preserving others, and it predicted that elementary particles would occur in a number of "generations." It turned out that the number of generations was far too big, but it may be possible to modify this first approach and add some new structure to the orbifold.

An alternative method is not to ask exactly how the

space is folded, but to begin by assuming that a compactified six-dimensional space *K* exists and then ask what properties it must have for physics to make sense.

We start with E8 × E8 in the ten-dimensional space, and we have to end up, at elementary particle distances, with something like an SU(3) × SU(2) × U(1) (grand unified) symmetry and things that look like elementary particles. Since the symmetries and properties of these particles are determined by what goes on in *K*, this means we already know quite a lot about what this compactified space must be like.

In the case of twistor space, it was possible to stick pins in each point of the space and then put a little arrow, or vector, on the top. Transformations within the twistor space could then be thought of in terms of what happens to these vectors. The same thing applies in *K*. When we want a space to have certain properties, we can sometimes express this in terms of how the arrows or vectors behave under transformations of the space. In the case of the compactified space, we need this structure to control the way in which E8 × E8 breaks down.

Knowing what symmetry we have to end up with has the effect of adding structure to the unknown compactified space *K*. It turns out that *K* must be the sort of space that was first studied by E. Calabi in the late 1950s and refined by S.T. Yau in the 1970s. The resulting space is therefore called a Calabi-Yau space.*

A Calabi-Yau space looks as if it will give the right symmetry breaking when a ten-dimensional space is rolled up. But the orbifold also looked as if it were on

---

*These Calabi-Yau spaces are cases of what are called Kahler manifolds. In attempts to go beyond the conventional space-time formulation of superstring theory and avoid working in a flat ten-dimensional space, some physicists have suggested working directly in these Kahler manifolds. For example, M. J. Bowick and Sarada Rajeev from MIT have created such a space out of the heterotic string loops themselves.

the right lines. Is there some connection between Calabi-Yau spaces and orbifolds?

In fact, an identification can indeed be made, but it requires some delicate space-time surgery. The problem is that the twenty-seven special, singular points in the orbifold mean that this space is not what mathematicians call a manifold. Manifolds can always be covered, like a patchwork quilt, by well-behaved coordinate systems. These twenty-seven special points prevent this happening in the case of an orbifold. But suppose we take a pair of mathematical tweezers and pick out each one of the twenty-seven points. We are now left with twenty-seven tiny "holes" in the space and, just as with the inner tube of a bicycle, we can now stick on twenty-seven patches. A little piece of complex space is glued over each of these holes in order to "mend" the space and convert it into a well-behaved manifold. Like an over-patched tire, the resulting space may not be smooth, but at least it is what mathematicians would call a topological manifold. In addition it continues to have all the properties we want, including a series of generations for the leptons and quarks.

What is the relationship of this patched-up orbifold to a Calabi-Yau space? It turns out that the Calabi-Yau space is a sort of smooth version of the repaired orbifold—it could be thought of as what happens when the radius of each of the twenty-seven patches shrinks to zero. We have therefore approached the same general space from two directions, one by looking at the actual geometrical ways in which a space could be compactified, the other by asking what sorts of structure that space should have in order to make sense. Certainly something of this approach must be on the right track.

Now that the two compactification approaches have linked up, what should be the next step? The actual examples we have looked at so far are not all that realistic. To begin with, thirty-six generations of leptons and quarks are far too many. Suppose, however, we generate some richer forms of geometry, just as earlier we moved from the simple torus to the orbifold. It turns out that there are a number of ways of doing this, and indeed some of them do end up with an account of the elementary particles in which four generations are predicted to be. But there are a variety of different ways of carrying out the whole compactification process, and what theoreticians really need is a compelling reason or principle to guide them through the compactified space and to push the theory in a unique direction. More and more of the physics of superstrings is becoming involved with this essential question of the geometry of the compactified space.

Beyond Perturbation Theory

String theory begins by writing down the relativistic equations for a single string and then quantizing this result. The interactions themselves emerge in a novel way in terms of the splitting and joining of strings. Yet the approach used to calculate all the quantities of interest in the theory still remains old-fashioned perturbation theory. Again and again we have met this perturbation idea, this philosophy of assuming a reasonable starting point and then trying to home in on the correct result by adding an infinite series of corrections. In the case of an electromagnetic field, these perturbation corrections were represented by Feynman diagrams, and, on summing up infinite numbers of terms, results of surprising accuracy were achieved. On the other hand, when it came to gravity, the perturbation series failed totally to represent a curved space-time by adding a finite number of corrections to an initially flat space-time.

There are always problems inherent in the perturbation approach. The first few terms of a series may be taking us to what we think is our target, yet when the full infinite terms are added, they may blow up to infinity or take us in some totally different direction.

Superstrings are supposed to be a deep theory about

gravity and matter. Yet physicists continue to treat them as if they exist in a flat background space-time, a picture they hope will be corrected by using a perturbation series. Clearly this whole approach is inadequate and obscures the whole power of the superstrings themselves. Penrose has showed how to develop a nonperturbative picture of gravitons, using his twistor approach. Some similarly novel approach may be needed for superstrings. Indeed, when a full, nonperturbative approach is finally developed, it may turn out that the entire superstring theory changes in radical ways, new features may appear, and we may discover how symmetry breaking actually takes place and how to calculate the masses of the elementary particles. Yet to go beyond the conventional perturbation approach is a profound step, and no one is quite sure how to make such a move. One thing, however, is clear: a nonperturbative approach demands the creation of a proper field theory for superstrings.

String Fields

The central issue of superstrings has become, in Edward Witten's opinion, how to understand the theory in a geometrical way. In the theory of relativity, for example, there is an elegant circle of ideas in which Einstein's underlying intuitions and concepts fit together naturally with the mathematics he employed. In creating his theory, Einstein was guided by his belief in the uniformity of nature. This means that while different observers may each see a particular event in their own particular way, they must eventually come up with the same basic laws of nature.

What is needed today, Witten argues, is a similar circle of physical and mathematical ideas involving superstrings, and this means an altogether more profound understanding of the theory. Once this has been achieved, it may well transform physics and even parts of mathematics in far-reaching ways.

How is it possible to develop this deeper coherence

between the ideas of superstring theory and their mathematical formulation? Witten feels that the deepest approach is in terms of a field theory. In addition, Schwarz, Green, and Gross have each written of the need to develop a proper field theory of superstrings. But how is this to be done?

Quantum field theory is a deeper way of dealing with quantum mechanical systems than the quantum theory of Heisenberg and Schrödinger. But the price of this depth and greater generality is the overall difficulty of the theory and certain formal problems that remain to be resolved. Quantum field theory was essentially the work of P. A. M. Dirac, who in 1927, two years after Heisenberg and Schrödinger had created quantum theory, showed how the new quantum concepts could be extended from atoms to the electromagnetic field. This field was treated as an "orchestra of oscillators." Where conventional quantum theory produces a wave function corresponding to each particular oscillation, the field theory creates a sort of super wave function composed out of all possible oscillator wave functions and goes on to quantize the excitations of this super wave function for the electromagnetic field.

Armed with a field theory, it was now possible to explain how an atom absorbs or emits radiation, for the quantum field theory was able to describe the birth and death of quantum oscillations within the electromagnetic field. Killing one level and creating another looks like the absorption and emission of photons—the quanta of vibrations of this field. The super wave function corresponding to the lowest energy level of the field, in which no excitations are present, is called the ground or vacuum state. Photons correspond to excitations out of this ground state and between the other states of the field.

Following Dirac's ground-breaking paper, physicists went on to apply the same technique to the electron. The quantum field for electrons is similarly a super wave function in which the birth and death of excitations correspond not to photons but to electrons. One way of generating the quantum field is using the individual wave functions generated by the Schrödinger theory as building blocks for the new quantum field. For this reason, the process is sometimes called *second quantization,* the Schrödinger approach being first quantization. Dirac was able to show that such a field theory produces excitations that represent not only particles but also antiparticles, that is, electrons with both negative and positive charges.

Field theory now dominates many areas of physics. For example, a metal's ability to conduct heat and electricity, reflect or absorb light, and so on is explained in terms of various excitations of quantum fields. The vibrations of the atomic lattice that occur when a metal is heated are now described as *phonons,* the quantum excitations of the underlying lattice field. It turns out that the excitations of the electron field can also interact via the phonon field. One implication of this phonon field interaction is that, under certain conditions, an attractive force binds the electron field together and produces the phenomenon of superconductivity.

Quantum field theory is also used in studying the elementary particles. The various advances outlined in Chapter 5 involve setting up a quantum field with all the necessary symmetries, e.g., grand unified symmetry, supersymmetry, and so on, then treating the elementary particles as the excitations of this quantum field. Today physicists are arguing that this approach must be applied to superstrings. Up to now only the first quantized approach has been used. Classical strings are created according to an action principle that is made manifestly covariant in order to produce minimal surfaces in space-time. (A new alternative would be to begin with general holomorphic curves in twistor space.) The various modes of vibration and rotation of these minimal surfaces are then quantized to produce string wave functions.

The proposed next step would be to generate, out of the wave functions for individual string vibrations and rotations, a super wave function for the string field. In the case of a heterotic string, the ground or vacuum state of this super wave function has E8 × E8 symmetry and must be written in a ten-dimensional space. However, as this space compactifies, the symmetry of the ground state of the quantum field would break, and the theory should end up describing excitations that have a grand unified symmetry. At distances that are large with respect to the superstring dimensions, this heterotic string field theory would then look like the quantum field for point particles, and its excitations would have the appearance of elementary particles.

In this mainstream approach, the superstring quantum field is written in space-time, admittedly a ten-dimensional space-time. Yet Green's heterotic string theory indicates that closed superstrings interact with the space in which they move and are inseparable from it. Space-time cannot be an inert background but must somehow be created out of the dynamics of strings themselves. This means that the string field theory should really be formulated in some deeper way. Possibly space-time and some new and modified form of the quantum theory would both emerge out of a more fundamental field theory. To make real progress, the whole field theory approach must be rethought in the context of the strings themselves. A number of physicists are currently pursuing this postmodern approach and analyzing just what a string field theory would entail.

Some of the deepest contributions in this area are coming from Edward Witten, who is at Princeton's Institute for Advanced Studies. In 1986 Witten created a particularly beautiful field theory that relies upon the branch of mathematics called cohomology to get away from the underlying idea of a geometry of points. Witten's theory has become a hallmark for string theoreticians; nevertheless it does contain problems, for it is a theory of open strings and does not therefore describe gravity. Remember that a string loop has the same quantum numbers as a graviton, and the way loops are born out of and die into the background suggests the way in which the graviton is an aspect of space-time structure. Witten's theory cannot therefore take account of this. On the other hand, open strings interact and therefore like to join their ends. But an open string whose ends join up becomes a closed string. Therefore a proper theory of open strings must also contain the possibility of closed loops. In its present form, Witten's string theory cannot take account of this phenomenon and is therefore limited.

But think what a theory of closed strings would mean. The closed loops are said to be the string forms of quantum gravitons and, from a distance or at the low-energy limit, these closed loops certainly look like gravitons. Yet, as soon as we move to energies and distances at which the actual string structure can be seen, it makes no sense to talk about gravitons. The quantum graviton is really an outmoded concept, something that belongs to a world that was envisioned before strings. Moreover it is meaningless to talk about these string loops as being "in" space-time. If space-time could really be treated as the background in which strings move and vibrate, then the very quantum fluctuations of such a continuous space-time would break it up into foam at exactly the same scale at which the strings themselves become important. Clearly the idea of a space-time background has to be abandoned, and a true closed string theory has to be formulated in a totally different way

Witten and others have pointed out that although superstrings are formulated in a ten-dimensional space-time, in another sense a string field is also a theory about two-dimensional surfaces. The fields are a quantum expression of all the dynamical movements of a world sheet. But these world sheets are two-dimensional; that is, they are created out of a one-dimensional line or loop stretched into the second dimension of time. In this sense a string field theory is therefore about the quantum properties of two-dimensional surfaces.

Considerable research is now being directed toward this approach. Witten has pointed out that the program can proceed in two directions. On one hand, physicists can study the topological properties of these surfaces and in this way produce powerful insights. But it is also possible to consider the geometry of these surfaces in another way. Think, for example, of the various relationships between the curves drawn on these surfaces. Taken together this set of relationships is a complementary way of understanding the surfaces and the corresponding field theory. But such a coherent set of relationships can be expressed algebraically. After all, algebra is about the manipulation of symbols, and these symbols can stand not only for numbers but for operations and relationships. In fact, we can go straight back to that basic insight of Descartes that geometry and algebra are complementary pictures, each producing its own sorts of insights.

So not only are the field theoreticians of superstring theory looking at the topological structures of the world surfaces of the string but also at their underlying algebra. It turns out that physicists had already for some time been looking at these branches of algebra (the sorts of algebras first discovered by William Kingdon Clifford and Hermann Günther Grassmann in the nineteenth century). Their idea was to discover ways of making quantum theory deeper and freeing it from its attachment to an underlying space-time. By dealing purely with algebraic structures, they believed, it should be possible to get rid of the underlying space-time structure and generalize quantum theory in a significant way. Now Witten and others are also exploring certain branches of algebra, in particular the Grassmann algebras, because of their relationship to a string field theory.

In the end, it may be necessary to go beyond even these approaches to string field theory and to explore new ideas. Witten, for example, believes that the twistor approach may provide some important clues. Suppose that we begin by working in twistor space. In the last chapter we learned how additional structure could be given to this space by adding what mathematicians call a fiber to each point to give the overall space a fiber bundle structure. Witten suggests that what should now be added is the strings themselves! Possibly a structure corresponding to an infinite number of classical strings would be found at each point in twistor space. This could well form the starting point of a superstring field theory.

Penrose for his part feels that this approach needs further thought. Essentially Witten works with what the Oxford group calls an ambitwistor space. On the one hand, this complex space has a built-in versatility; on the other hand, this freedom is gained at the expense of a certain subtlety that is inherent in the initial twistor picture. Take, for example, the meaning of a photon in the twistor picture. Photons, the quantum particles of light, have an associated spin, which can point in either of two directions. It turns out that, in the twistor picture, a photon spinning in one direction is represented by a twistor function of homogeneity 2 (that is, a function containing terms like x^2, xy, and y^2), while a photon that spins in the opposite direction is represented by a function of homogeneity −4 (that is, a function containing terms like $\frac{1}{x^4}$, $\frac{1}{x^3y}$, $\frac{1}{x^2y^2}$, etc.). So a general quantum mechanical photon, which must be represented as having a combination of both spins at the same time, is therefore represented by a function of mixed homogeneity. This is a very curious result, yet one that Penrose feels correctly reflects the quantum strangeness of the world. Likewise, in the case of gravitons, the theory must express combinations of complex spaces that curve in left- and in right-handed ways.

But Witten's approach, while it is immediately more tractable, has also sacrificed this important quantum aspect of the earlier twistor approach. Rather than using two functions with very different homogeneity, Witten uses a more ambidextrous approach, which is in a sense closer to a quantum than a classical description.

Daniel Friedan and Stephen Shenker, both of the physics department at the University of Chicago, are pursuing their own unorthodox approach to the field theory of strings. Like Roger Penrose, they lay great stress on conformal invariance (that is, the symmetry of a massless, scaleless universe in which the structure of the light cone is unchanged by conformal transformations) as a key to creating a field theory. Their goal is to develop a deeper approach to superstrings, and that means to get away from the perturbation approach—the adding up of tiny corrections—and the whole space-time picture. Therefore they begin in a more abstract way without any reference to a background space-time. Since superstrings, in the closed-loop picture, give an intuitive picture of the gravitational force, why not let space-time emerge in a dynamic way from an underlying theory, using only the strings themselves?

The abstract space created by their strings is quite complicated. Its internal structure is governed by conformal invariance, but this space also has an infinite genus—the genus of a space is a topological property, and this space could be thought of as one with an infinite number of handles!

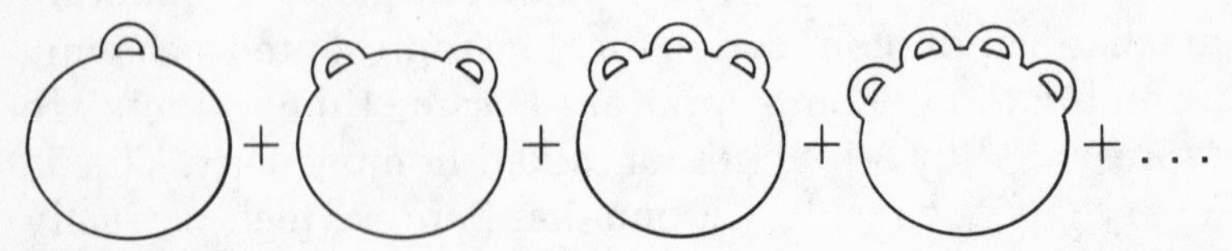

Figure 10–6
An infinite genus space is created by summing spaces with all possible numbers of handles.

Some physicists would argue that this approach gets away from perturbation theory because a space with an infinite number of handles could also be thought of as containing an infinite number of strings. So, in this sense, theoreticians are working in a space with an infinite number of strings and are getting away from the sort of approach that adds progressive perturbation corrections to some initial state containing a single string.

Friedan and Shenker believe that it will be possible to create a much deeper string field theory by beginning with this abstract space. They have developed something that looks like a wave function; however, it differs in its implications from the usual quantum theory. Like Penrose, they feel that a new theory of space-time must also generate a more fundamental quantum theory. In essence, by allowing space-time and quantum theory to emerge out of the one underlying approach, they are trying to create physics in a new way. Of course, all this is ongoing research, and it remains to be seen how their ideas will develop.

Another radical approach to string field theory is being carried out by M. J. Bowick and Sarada Rajeev at MIT. As does Edward Witten, Bowick and Rajeev point to the way in which Einstein's physical intuitions and arguments found their natural expression in the curved geometry of Riemann. String theory also requires a natural space for its proper formulation, they argue. For the MIT group this natural space should be generated by the strings themselves.

They begin with closed-loop strings, along with all their excitations, and, using this structure, go on to create a complex space of infinite dimensions—called a Kahler geometry. (We in fact met this Kahler space before in connection with the question of compactification.) The philosophy of this approach is not unlike that of twistor theory. Reversing the argument of Chapter 7, the twistors, created to have both angular and linear momentum, can be thought of as null lines and congruences of null lines in space-time. Yet out of their space-time relationships, it is possible to create a more fundamental twistor space. The points of this twistor space are generated by the twistors and their congruences in space-time.

In a similar way, Bowick and Rajeev use the strings and all their excitations to create a space, this time with an infinite number of complex dimensions. The next step is to develop a superstring theory in terms of the geometrical structure of this space. At atomic distances the superstring loops look like dimensionless points, and at this scale the complex Kahler space must therefore reduce to ordinary curved four-dimensional space-time. In fact, Bowick and Rajeev are able to derive an equation that is analogous to Einstein's field equation in a curved Riemannian space-time. Again, this is a highly speculative approach that, at this stage, looks as if it could be along the right lines and could result in all of physics being derived at a much more fundamental level.

Stop the Presses: Shortly after the manuscript for this book had gone to the publisher, I learned that Edward Witten had come up with a totally new field theory. Indeed, he had created a great deal of excitement at Oxford University by lecturing there about his new ideas in March 1988. In essence, the new approach is a very powerful, yet very abstract field theory. While superstrings have been hailed as the theory of everything, Witten's new approach could be said to be a theory of nothing, for it is really a field theory about some minimal but basic mathematical structures. Some physicists feel that this approach may give clues as to the correct way to describe nature and distances below the Planck length—distances at which the whole idea of space-time breaks down.

Below the Planck length, we would not expect to see many of the usual space-time properties; there would be no sense of length, no measure or metric to the space. In fact, the "space-time" would be very abstract and general; then, at around the Planck length, something like a phase transition occurs. (A phase transition takes place when ice changes into liquid water, or water into steam. While a dramatic change of order occurs at 0°C or 100°C, the "essence" of water in its molecular form does not, however, change, so that "water" could be said to exist in several phases or forms.) At the Planck length we jump from a space without metric to one with a particular measure. Above the Planck length distance has a meaning; below the Planck length physicists would have to rely upon topological properties.

One way of thinking about Witten's abstract new theory is that the universe we live in actually breaks nature's deepest symmetry. Einstein taught that all theories must be manifestly covariant (as is Witten's), which means that any coordinate system is as good as any other. But it is possible to argue that our own universe has singled out a particular solution, a particular coordinate description, in which space-time is essentially flat. Of course, it is true that the presence of matter, such as stars and planets, acts to curve the geometry of space-time, but on average our space-time is flat. This could be taken to mean that out of all possible geometries, our own universe has selected just one and thereby broken the underlying symmetry of nature, the same symmetry that is present in Witten's theory.

Suppose that the universe was created in a highly symmetric form without metric or measure. At some point, it then undergoes a phase change and collapses into the particular space-time around us. In terms of string theory, this means that a specific solution, a particular form of string theory, must likewise condense out of Witten's more general description. So particular string theories are in fact the particular phases that condense out of Witten's new field theory.

A striking aspect of Witten's new approach is that it is formulated not in ten dimensions but in four! Does this mean that superstring theorists are abandoning their flirtation with higher-dimensional spaces and returning home to a more familiar four-dimensional space-time? In fact, the Oxford mathematician S. Donaldson has demonstrated that some powerful mathematical structures are unique to four-dimensional spaces. If physicists want to use such special mathematics, then they must abandon their higher-dimensional space-times and work in four dimensions.

Light Cone Gauge

Superstring theory has a guilty secret, which will now be revealed! In a sense, everything that has been done up to now has been a bit of a cheat, for while the superstrings were initially set up to be manifestly covariant (and satisfy the beautiful general symmetries of general relativity), in practice, physicists have then taken a simpler route and in the process have obscured much of the power that underlies the full theory.

In defining a problem in physics, it is usual to give what is called initial data. We begin by writing down everything relevant that is known about the system. A shot in pool is described by giving the position of the stationary colored balls plus the position and momentum of the moving cue ball. Then, armed with Newton's laws, it is possible to compute how all the collisions occur and which balls fall into which pockets.

In relativity theory things are done in a more complicated yet more symmetrical way. The initial data of the pool game was given at all points in space and for a single instant of time. This does not make physical sense in relativity—the idea that information can be collected from many distant points all at the same instant of time implies that this information must travel across space at infinite speed—faster than the speed of light. Rather than specifying data at a single instant, physicists must give all the data that lie on a slice of space-time. (We met this idea of slices through space-time in Chapter 6, where we discussed Penrose's approach to field theory.)

But which slice is to be taken? You and I move at different speeds through space-time and agree to specify our initial data on a slice. But your slice will be at a different angle from mine. How are we to come to any agreement? The answer lies in the covariance of Einstein's theory. General relativity says that the form of a physical law must not change when we move from observer to observer, from coordinate system to coordinate system. This means that the laws must be independent of the choice of space-time slice, or space-time gauge as it is called.

Superstrings were created to be covariant, but this also means that when calculations are made in the theory, for example, the sorts of interactions represented by pairs of trousers, then it is very important that these calculations should not be tied to any particular gauge, or choice of space-time slice. The guilty secret associated with superstrings is that physicists have found this too difficult in practice. While the theory is created in a covariant way, the actual calculations are always made using a particular convention for taking space-time slices called the *light cone gauge*. In this sense, the whole theory is limited.

The light cone gauge could be thought of as a sort of region of space or a three-dimensional surface that moves through space-time at the speed of light—what relativists would call a "null hyperplane." The initial conditions corresponding to a particular physical situation are defined on this surface, and the actual system is then followed as it evolves within the whole of space-time. In fact, everything that has been said about superstrings in the previous chapters has been done in this particular light cone gauge. For example, the pairs of trousers, along with the holes in their waistbands, only make sense in the light cone gauge.

But although this choice of gauge makes physical sense, it violates the full relativistic covariance that is inherent in superstring theory and decrees that the details of the theory should not really depend on the particular choice of gauge. By obscuring this underlying symmetry, physicists are really working with the shadow of a much deeper theory. It also turns out that using this single gauge is intimately connected to perturbation theory along with all its limitations. One of the important tasks of this postmodern physics is to rewrite string theory so that it is free from any particular choice of gauge or space-time slice. In this way, it may be possible to penetrate much deeper into the theory and resolve some of its present difficulties.

How Many String Theories Are There?

The crisis that was facing elementary particle physics at the end of the 1970s was the multiplicity and arbitrariness of the various theories, their modifications and variations. When it came to grouping the elementary particles into patterns, for example, there seemed to be no definitive way of choosing between the various schemes. Then Green and Schwarz came along and proposed a unique theory, a theory of supersymmetric strings that was not only free from all the sorts of problems that had plagued point field theories but one that would work only with a unique choice of symmetry. The great virtue of superstrings, it appeared, lay in their uniqueness.

Today there are a variety of string theories, not only those of Green and Schwarz and the heterotic strings of Gross, but newer theories, variations, and modifications. Today string theories may well run into the thousands. What then has happened to their uniqueness?

In this sense, things look bleak for superstrings. However, a new philosophy is emerging which argues that, at its deepest level, string theory *is* unique, and the various theories we see around today are simply different *phases* of the one, unified theory. How can this be true? A good example is given by the element carbon. As an element, carbon is unique—there is only one sort of carbon atom*—yet carbon itself can be found in several different forms, as charcoal, graphite, or diamond. While

*The fact that the carbon atom can exist in different isotopic forms does not affect this argument.

charcoal is black and crumbly, diamond is clear and transparent—the hardest natural substance known. Graphite, while it also has a crystalline form, is black and very soft; indeed it can be used as a lubricant. While there is only one carbon atom, only a single element, nevertheless this element can appear in a variety of quite different forms or phases.

Some physicists are now suggesting that, by analogy to carbon, there is only one underlying string theory, but this basic theory can appear in a wide variety of different phases. While the various formulations of string theory are therefore perfectly correct, they are *only particular* solutions of the more general underlying theory.

One suggestion is that the universe began in a unique way, with this single description in terms of superstrings. But then, as with carbon, it evolved into one of a number of alternative string phases. Our universe just happens to occupy one out of a number of possible phases or forms.

But what is the nature of this initial form? To talk about such an arena of physics will probably require some new key ideas. Again and again we have emphasized that it makes no sense to talk about a string vibrating *in* space-time; essentially the string and space-time at this level must be inseparable—they are one and the same thing. The whole idea of a background space-time has to be abandoned.

Some physicists feel that at its most fundamental, string theory is really about a two-dimensional world sheet. But even the ideas of dimension themselves, of distance and space-time structure, may not make sense at this deepest of levels. It could be that what is called for is a sort of prespace description for the true origin of a string world. Then, at some particular point in . . . but a particular point in what? Certainly not space or time, for these are now classical, large-scale things. So let us say at some point—in something—there is a sudden change, in which space is born and along with it one of the particular phases or versions of superstring theory. Current thinking is that this birth of space must happen around the Planck length. Remember that some of the early arguments about space-time structure suggested that the fabric of space-time breaks into a foamlike structure below the Planck length.

But what would a theory of prespace really look like? In a sense, this question brings us right back to the sort of issues that have preoccupied, not the mainstream physics community, but those few mavericks who could never be content with the quantum theory and general relativity as they stood in splendid isolation from each other. Roger Penrose has spent over twenty years trying to work toward such a theory. Michael Green, for his part, argues that a theory of prespace would mean letting go of so many of our familiar, classical ideas. And, of course, so many of these are still based upon the notions of space and time. To suggest that space-time and strings emerge out of a much deeper theory is one thing, but to say just what such a prespace theory means is quite another. String physicists today face a major challenge.

The first chapter of this book pointed out how pervasive is the Cartesian order with its ideas of continuity and the dimensionless point. Yet this same Cartesian order is clearly incompatible with the meaning of the quantum theory. String theory looked like a significant first step in freeing physics from these classical, Cartesian notions. Nevertheless they still continue to exert a tremendous hold on our thinking; so much of mathematics, for example, has evolved from the assumptions of continuity. Even though string physicists are looking for new ways of describing extended objects, they are again and again falling back upon the use of such powerful tools as continuous field theories, smooth manifolds, and the use of the calculus.

The calculus and all the mathematics related to it have achieved miraculous results in theoretical physics over the last 300 years. It is therefore difficult for physicists to give up such a powerful tool, and along with it the idea of what is called a differentiable manifold—that is, a space in which the calculus is possible. On the other hand, how is it possible for such ideas to persist below the Planck length? Twistor theory, for example, does not accept that space-time need be smooth or that we should expect such notions to continue to be useful at the subatomic level. Penrose's spin network was an attempt to derive the properties of a space that was essentially quantum mechanical. It could well be that an extension of this idea is possible within twistor theory itself. Penrose is quite happy to abandon the notions of continuity when it comes to space; however, it is clear that he would still like to hang on to some sort of notion of complex analycity and the power that comes from the mathematics of complex numbers.

The implications of string theory and twistors are that old ideas and habits of thought must be swept away and that the revolution in physics that began with relativity and quantum theory has yet to be fully achieved.

What Is the Dimensionality of the Universe?

The whole question of dimensionality has become the burning issue of string theory. In its first appearance, as formulated by Nambu, string theory called for twenty-six dimensions. The superstrings of Green and Schwarz required only ten dimensions, and then Gross came along with his curious heterotic mixture of dimensions running in different directions along a closed string—a theory that contained both ten and twenty-six dimensions. More recently string physicists have given attention to the world sheet of the string itself, which is a two-dimensional object. In one sense all the physics could be said to arise in the vibrations and quantum fluctuations of this string. But does this really mean that string theory is neither twenty-six- nor ten-dimensional but really two-dimensional?

A physicist like Abdus Salam—creator with Weinberg of the unified electroweak theory—believes that indeed two dimensions are the most fundamental. "God created two dimensions," he says. In other words, physics is born in the two dimensions of the string's world sheet. At some later point, a transition occurs to ten dimensions, and finally this larger space compactifies down to our more familiar four dimensions. According to this idea, the dimensions of the universe are not immutable and fixed but evolve from a more fundamental two-dimensional state.

In Green and Schwarz's first formulation, ten dimensions seemed to be the inevitable way of doing things. Once that road had been taken, then physicists next had to go on to discuss ways of getting rid of six of these dimensions through the mechanism of compactification. It was almost possible to become convinced that ten dimensions were indeed the natural way to describe the universe, with our four dimensions of space-time emerging as a sort of afterthought. But theoreticians are now questioning whether all these theories really have to be formulated in ten dimensions. Or was this simply an assumption that happened to develop historically because, in the early forms of the theory, ten dimensions looked like the best way of proceeding? Even back in the early 1980s, the famous Russian physicist A. M. Polyakov was working in spaces of fewer dimensions. Indeed his particular approach, which involves working in a lower-dimensional space, seems to be leading physicists to new insights on Nambu's original string theory!

Other string physicists are now beginning to suggest that it is possible these additional dimensions are, in Michael Green's words, "not really dimensions at all." Green and Schwarz's original theory could be thought of as involving a four-dimensional space-time plus six additional "things" that may not be dimensions as we know them. Likewise, Gross's heterotic string theory may contain a combination of a four-dimensional space-time plus six "things" moving in one direction (for the fermions) and twenty-two moving in another (for the bosons).

But if these extra dimensions are not really dimensions at all, then what are they supposed to be? At the present state of the game, it is difficult to get a clear answer, because the underlying mathematics has yet to be worked out. One approach would be to say that dimensions, as we know them, tend to be large-scale things, properties of our own macroscopic world. But when one deals with the strings themselves, it is possible that new concepts and ideas are required. It is almost as if in using terms like *dimensions* physicists were trying to hang on to old-fashioned ideas that no longer apply within the world of strings.

In fact, the idea of dimension has already received a setback from quite a different field—the study of fractals. The mathematician Benoit Mandelbrot has been highly active in the study of these curious, infinitely detailed figures which, among other things, have fractional dimensions. According to Mandelbrot, the most interesting things in our world—clouds, mountains, coastlines, lungs, trees, nervous systems, and galaxies—do not have simple dimensions but, rather, fractional dimensions. In fact, up to now geometry has been dominated by those rather dull one-, two-, and three-dimensional figures first studied by the Greeks. But in Mandelbrot's world, the dimension of a figure changes as we move toward it and see it in ever greater detail. Dimensions are no longer fixed but behave in curious mutable ways.

To my knowledge, there is no direct connection between Mandelbrot's fractals and the dimensions of superstring theory. The purpose of this example was simply to shake your faith in the classical idea of dimensionality. Possibly dimensions are not as obvious as everyone thought and, at the scale of superstrings, some totally new concept is needed. Once we leave the world of point particle theories behind, it may be necessary to put on a new set of intellectual clothing.

Superstrings and the Aging of the Universe

Like any new theory, superstrings have a number of curious and unexpected spin-offs. One of these concerns a new particle called the dilaton, predicted by the theory, and its connection to the age of the universe. The great English physicist P. A. M. Dirac, who we met earlier as the inventor of quantum field theory and the fundamental symmetry between particles and antiparticles, was greatly struck by the coincidences between the values of certain fundamental numbers that appear in physics.

Dirac noted that the ratio of the strength of the electrical force to the strength of the gravitational force can be expressed in terms of the fundamental constants of nature. The answer is an extremely large number—10^{39}. This number does not depend upon the choice of any particular units or scale. Just as the ratio of a person's height to the length of his or her arms is independent of the particular units of length, so too this ratio of fundamental forces is independent of the choice of units.

Dirac also noted that the number of elementary particles in the universe—again a dimensionless number that can be expressed in terms of the fundamental constants—is 10^{79} which is approximately the square of the earlier number. This was puzzling, since all the other dimensionless numbers of physics are small. Even more puzzling was the coincidence with the age of the universe, which, when expressed in appropriate atomic units, is also 10^{39}. Was this simply an accident? Dirac's keen aesthetic sense would not admit such coincidence in the planning of the universe. But this must mean that the ratio of the fundamental forces, the ratios of certain constants of nature, and the number of particles in the universe are all related to the age of the universe. If this is true, then certain of the fundamental constants must change with time!

One implication of this theory is that the gravitational constant (which is a measure of the force with which the earth attracts apples and the moon, and the sun and earth pull toward each other) is growing weaker. Another implication is a tiny, gradual change in the value of the speed of light. In addition, the conservation of mass and energy breaks down because new elementary particles are constantly being added to the universe. By combining the ideas of continuous creation of new matter and a much greater gravitational pull in the past, Dirac deduced that the earth would have had a much smaller radius in remote geological eras and has since been in a constant state of expansion. Some people have even tried to tie this hypothesis of an expanding earth to the theory of continental drift. In essence, the protocontinents would have covered a much smaller earth and, as a result of the earth's expansion, they have since drifted across its surface.

But very careful measurements suggest that if the gravitational constant is indeed changing, then it cannot be as fast as is suggested by Dirac's theory. In fact, the whole idea has not been taken seriously over the last decade. However, with the development of superstrings, the possibility that the fundamental constants of nature are varying with time has taken on a new significance. In superstring theory, a particle is predicted called the *dilaton*. According to theory, the dilaton should appear in a massless form. But a massless dilaton would have the immediate physical consequence that the constants of nature must change with time—in fact, a revival of Dirac's basic intuition. Many physicists feel that these temporal changes are ruled out by experiment, and this means that the dilaton, if it does exist, must somehow obtain a mass. Thinking up ways in which particles can gain mass is not too difficult where symmetry breaking is concerned. But if the dilaton is indeed given a mass, then this creates other problems, for the particle will then exert what looks like an additional gravitational effect and produce deviations from the famous inverse square law of force that was first established 300 years ago by Newton.

The existence of the dilaton gives theoreticians two options; either look for a correction to the established law of gravitational interaction or accept the aging of certain fundamental constants with time. At present the issue is still an open question. Of course, superstring theory itself suggests that there must be fundamental changes to Einstein's theory of gravity, for the interactions between matter and a curved space-time emerge out of a much deeper theory in which loops are born or die into space-time.

In fact, it turns out that the law of gravity is under close experimental scrutiny at present. The previous chapter pointed out that one of the significant clues that led Einstein to his general theory of relativity was the equivalence between gravitational and internal mass. This equivalence, first noted by Galileo, was confirmed by the Hungarian physicist Roland von Eötvös at the turn of the century.

Recently Ephraim Fischbach of Purdue has analyzed Eötvös's experimental data and argued that the tiny inconsistencies that Eötvös believed were random experimental errors are so systematic as to point to an additional repulsive force. If Fischbach is correct, then this would mean that a fifth force of nature had been discovered which acts over several hundred yards and is repulsive rather than attractive.

A number of very precise experiments have since been carried out to test this idea. The results, however, are controversial and inconsistent. Some groups have discovered no evidence for a fifth force; others have found a small additional repulsion that appears to depend upon the actual composition of the material involved, for example, the total number of protons and neutrons or the total isospin. At all events, the gravitational force looks as if it will be coming under close and accurate scrutiny, both theoretically and experimentally, over the next few years.

Cosmic Strings

When I was a boy, one of the experiments—carried out in my laboratory located in a coal shed—that used to turn me on involved a supersaturated solution of sodium thiosulfate. It appears that the phenomenon I was observing all those years ago is not unconnected to the idea of those mysterious cosmic strings that were proposed just over ten years ago and may even have some connection to superstrings.

A considerable quantity of sodium thiosulfate, or "hypo" for short, can be dissolved in hot water. This solution does not crystallize out on cooling; rather, it forms what is called a "supersaturated solution." However, if a single crystal of hypo is then dropped into this supersaturated solution, the whole thing begins to crystallize in an almost explosive fashion. Crystals stretch out from all over the liquid, and within seconds it becomes a solid crystalline mass. Something analogous may well have happened within the first second of the big bang creation of the universe.

The saturated solution is highly unstable; it contains too much hypo, and its energy is therefore far too high. Yet, like a roller coaster car that teeters at the top of the hill, it is stuck in the highly unstable position. Add a single crystal, however, and this acts as a nucleus for crystallization. Suddenly the unstable solution crystallizes, and in this way the energy drops to a stable level.

In a similar fashion, it is believed that the early universe existed in a highly energetic yet symmetric state. Like the supersaturated solution, it too was trapped on the top of a hill of energy. It could only gain stability by breaking its high degree of symmetry, in which all the forces of nature are unified. This symmetry breaking of the universe is assumed to have all taken place long before the first second of the creation of time had passed. The result was the separation of the gluon, electromagnetic, weak, and gravitational forces and the appearances of elementary particles with their unique pattern of masses.

When a supersaturated solution of hypo crystallizes, the various crystals grow extremely rapidly and bump into each other so that it is possible to have regions of

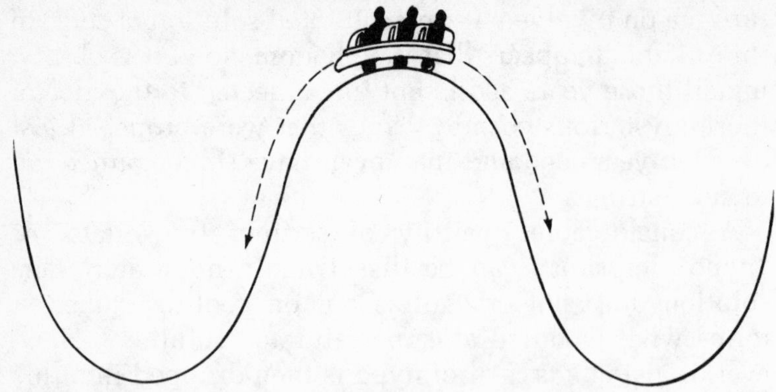

Figure 10–7
A universe could exist in an unstable state, poised upon a hill between two energy valleys. By choosing to fall into one valley and release its surplus energy, the universe will also break a basic symmetry. (Clearly the case where the roller coaster lies in the left or the right valley is less symmetrical than when it is midway between them.)

supersaturated solution trapped by various defects and boundaries between the crystals. The apparently solid crystalline mass may contain within its defects tiny regions of high-energy, supersaturated solution. Something similar is also believed to have happened during the "crystallization" of the very early universe. So within our universe it is thought to be possible to have trapped regions of extremely high energy corresponding to the initial highly symmetric state of the universe.

In 1976 T. W. B. Kibble of Imperial College in London proposed that such defects, called cosmic strings, were indeed formed during the early stages of the creation of the universe. These strings could exist as closed loops or be of infinite length. Their thickness would be a mere 10^{-30} cm, yet such an immense amount of energy from the primordial universe is still trapped inside them that one inch of such a string would weigh 10 million billion tons! Stretching across the universe, these strings would curve space-time and exert a considerable gravitational attraction over matter in the universe. In the early 1980s, Y. B. Zel'dovich of the Institute of Physical Problems in Moscow argued that such strings could explain the curious clumping together of matter on a cosmic scale.

Edward Witten has shown that cosmic strings can exist in the superconducting state, carrying electrical currents as high as 10 million billion amperes. Since the strings also vibrate, these enormous electrical currents will oscillate and emit extremely powerful bursts of radiation. Radiation that intense would burst out across space with such force that it could push away the surrounding matter and, as it were, blow a cosmic bubble in space. Some evidence for such cosmic bubbles was discovered in 1986. Mapping of nearby parts of the universe indicates that galaxies look as if they are forming around the edges of cosmic bubbles. If such immense bursts of energy are still occurring, they could also be detected as bursts of x-rays that appear to come from ring formations far away in space. As these rings collapse, they would release intense bursts of elementary particles, which could explain the origins of the highest-energy cosmic rays. At the low-energy end, it is possible that the thread-like radio sources that have also been discovered at the center of the Milky Way could be produced from low-energy cosmic strings.

Cosmic strings are, at present, a purely hypothetical construct that make a number of curious and dramatic predictions about the large-scale structure of the universe. In addition, they suggest such things as the bending of light to form double images around a string, together with a number of other possibilities.

And what, if anything, do cosmic strings have to do with superstrings? Maybe there is no real connection; however, Hughston and Shaw believe that the proper way to understand these cosmic strings is to begin in twistor space so that the cosmic strings emerge as spacetime structures corresponding to holomorphic twistor

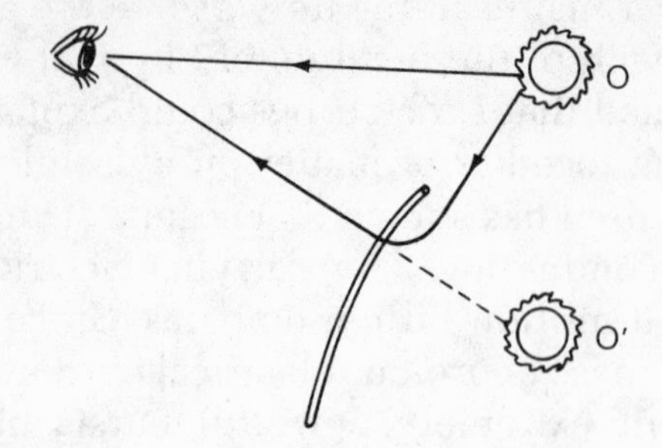

Figure 10–8
Light from a distant astronomical object reaches the eye in two ways, directly, and after being bent by the intensely curved space-time around a cosmic string. The light that has been bent around the cosmic string appears to have come from a different location in space and therefore results in a second image *O′*. In a sense, the cosmic string produces a mirage in outer space.

curves. In other words, both the relativistic string—the precursor of a superstring— and the cosmic string emerge as being different solutions to an equation first proposed by Nambu. In itself, this does not really mean that superstrings and cosmic strings have any deeper connection. Indeed that there could be any connection between the very largest and the very smallest structures in the universe would make even the most liberal of minds boggle.

Knots in the Superstring

In 1984 superstrings were being hailed as the "theory of everything." But now that the first flush of their success is over, physicists are beginning to add up the account and weigh the credits against the debits. The credits are certainly impressive. John Schwarz has termed some of these "miracles"—for example, that a consistent theory exists that is compatible with quantum theory, relativity, and causality; that the theory incorporates gravity in such a profound way; and that it is able to explain so many features of the standard model of the elementary particles.

On the other hand, there are certainly a number of doubts and questions raised by superstrings. Let us look briefly at some of them:

- To begin with, there is a vast gap in energy between superstrings and elementary particles, so large indeed that everything that physics has studied up to now is nothing more than a hyperfine correction to superstrings. Is it going to be possible to derive all this additional fine structure out of the superstring theory?
- The appeal of Green and Schwarz's initial paper lay in its uniqueness. In contrast to the many attempts at a grand unified, supersymmetric theory of point particles, superstrings emerged with a unique symmetry. Then came the alternative formulation of heterotic string theory. Today it is not clear whether there is a single unique string theory or up to a thousand variants! Is string theory going the same route as the earlier grand unified theories? Physicists would prefer the inevitability of a single theory.
- There must be a compelling reason why the ten-dimensional space-time compactifies down to our four-dimensional space plus a compactified space *K*. In a truly deep theory, the dynamics of compactification and the properties of this compact space should be determined. It should also be clear that this compactification is stable and the curled-up dimensions never change in radius.
- Why exactly is the length of superstrings fixed to be equal to the Planck length—why not 100 or 1,000 times that length? Of course, it is a neat thing to fix the length of superstrings at exactly the length at which space-time structure begins to break down, but is there something compelling within superstring theory itself that fixes this length?
- Connected to this whole question of compactification is the vanishing of the cosmological constant. Why

is it that we live in a four-dimensional space-time that is flat in the absence of matter?

- What is the essential physics of the string world? How do things behave at really high energies? While such energies are beyond anything that we could ever hope to measure in the laboratory, there could be aspects of string theory that have cosmological significance or were important during the creation of the universe.

- Is there some way in which the theory of superstrings can be tested? It "retrodicts" a number of things; that is, it makes predictions of things that are already well known. It also predicts a number of things that can probably be predicted by more orthodox approaches. But is there some measurable result that is essential to a string theory that would not be expected on the basis of any other approach?

The late Richard Feynman, for example, has argued that superstrings don't really calculate anything at all, that they simply cook things up and contain too many arbitrary elements. Other physicists have speculated that it may turn out to be technically too difficult to predict new things with superstring theory and that the theory could atrophy because of this lack of predictability.

Of course, other physicists hope that new experiments done with bigger and better elementary particle accelerators may one day be able to confirm a prediction of superstring theory. But here again Feynman makes a very interesting point. We don't need new experiments, he argues, for we have enough unexplained things around already. Take, for example, the masses of the elementary particles. These are very well known, but no theory has been able to explain their particular values. If string theory is truly a theory of everything, then its greatest success could come from just this, from giving an account of all the masses of the elementary particles. Unfortunately the current string theories can give no unique explanation for these values.

These are all practical, scientific questions, but there is another way in which the superstring approach is being criticized. In researching this book, I have discovered that a number of eminent and respected physicists have strong reservations about the whole superstring approach. Their doubts are not about technical matters so much as a criticism of the sociological implications of the whole enterprise in which vast numbers of scientists and their students have ended up all working on the same field.

The most talented of theoreticians are drawn to superstrings by their challenge, others by the glamor of the topic or by the demands of their students. After all, students majoring in physics want to discover what all the fuss is about, and a graduate student with an eye on a job in the near future feels that some research experience in superstrings would be of advantage. The result is that an extremely large number of theoretical physicists are all crammed into the one field. The problem is not so much that this theory could turn out to be wrong, but rather that it does not seem proper for so many scientists to be all going in the same direction at once.

In fact, it is not difficult to see how this concentration of brain power came about. In the sixties and seventies, a number of different approaches were at the frontiers of physics. People were looking at string theory, grand unified theories, supersymmetry, and the Kaluza-Klein approach. But the anomalies and dead ends associated with these theories produced a sense of disappointment. Then, when the superstring breakthrough came along, scientists from these different approaches suddenly realized that they had something in common. It was possible to bring their special talents to bear on this new field, and—at first at least—it looked as if there would be no need to learn a lot of difficult new mathematics. Scientists and research groups with nowhere else to go were willing to jump onto the superstring bandwagon.

Nevertheless some physicists are uneasy, for they feel that the whole superstring approach has been overinflated and that other avenues of research should also be explored. Yet superstrings have the glamor, and as more and more articles are published, the whole business becomes self-fulfilling—if so many people are writing about the theory, then there must be something in it!

Of course, the same criticism could have been applied in 1925—everyone jumped on the bandwagon of the new quantum theory, and who would now object that this was not a smart thing to do? All the same, many of the leading physicists of that time, including Einstein, de Broglie, Schrödinger, Lande, and others, voiced serious doubts, and their criticism helped to sharpen up the final theory. Today we are beginning to acknowledge that there are serious problems with the interpretation of quantum theory and the foundation of quantum field theory. Who knows whether things would not be clearer today if some scientists in the 1920s had chosen to pursue alternative ideas and theories? In retrospect, it is very difficult to judge.

But as far as the superstring critics are concerned, physics today should be exhibiting a healthier divergence of ideas. Other avenues should be opening up and alternative theories developed. With a diversity of new ideas, the next step would be for scientists to begin to look for underlying connections, or the ways in which the same phenomena could be explained in different ways. In this fashion the whole enterprise becomes clearer and deeper. It is certainly true, for example, that superstring theory has benefited from its connection to twistors. Penrose and his twistor group avoided the elementary particle bandwagon, and for a while it may have appeared that their work had little relevance to mainstream physics. Nevertheless, deep and powerful connections have now been made to superstrings.

There is another respect in which the superstring approach is being criticized, and that is because of its excessive mathematical content. Michael Green, one of its creators, feels that the growth of abstract mathematics within the theory is very exciting. Nevertheless he is concerned that a deep physical principle is still lacking that would guide the development of the theory. Such a principle, he argues, would revolutionize our understanding of physics. David Bohm's objections to the increasing mathematization of physics can be found in his dialogue with me which opens our book *Science, Order and Creativity* (New York: Bantam New Age Books, 1987).

Bohm feels that physics has relied excessively on mathematics during the last half-century, to the point where physicists are ignoring the philosophical and physical underpinnings of their theories and concentrating on the mathematics. For Bohm, mathematics is a relatively limited language in which to understand the universe. His objections to the superstring approach are to what he feels is its ad hoc nature and the absence of a fundamental guiding principle.

On the other hand, many physicists are happy to engage in the unfolding of superstrings and feel that, as new connections are made, a deeper principle will eventually emerge. But what sort of a principle could this be?

The Search for a Deeper Principle

A truly fundamental linking of twistors to superstrings would mean that not only space-time would be generated out of this deeper theory but a new form of quantum theory as well. Present day quantum theory would then become an approximation to some much deeper theory, and with luck, the new version may avoid some of the outstanding problems of the present theory.

A major problem that would also have to be resolved would be the meaning of compactification. Are ten dimensions the proper starting point for physics, or should it be a complex twistor space? Present attempts to link twistors to superstrings involve attempts to join the two theories, but it is also possible that essential features of each could emerge out of some even deeper theory.

It is clear that a deeper principle is required to guide physicists toward some extended theory. But what is the nature of this principle? On the basis of what we now know about superstrings, it should involve some overall insights about the theory's underlying cohomological structure. Essentially some new understanding of the basic uniformity and economy of nature may be connected to some deep structure of the superstring space. It could turn out, for example, that nature is unchanged under certain transformations of structures in a fundamental space. By relating these structures, together we should then be able to recover certain laws. For example, Penrose's twistor approach suggests that this basic cohomology and the preservation of certain structures in twistor space may be more fundamental than the action principles that have been traditionally used in both classical and quantum physics.

Another hint as to this guiding principle may come from notions of symmetry and order. Majority opinion holds that nature in its most fundamental state is highly ordered and symmetrical so that its underlying laws are mathematically simple and unified. Then, through a sequence of symmetry-breaking processes, we end up with the particular complexity we see around us, and with laws whose validity holds only within more limited contexts. Penrose, however, has attacked this basic philosophy by suggesting, with the help of his graviton example, that nature begins in a more asymmetric way and that some of this asymmetry is then averaged out.

Clearly there is both symmetry and asymmetry in the universe. But which is more fundamental? Should we begin with simple order and symmetry and then account for its breaking? Or is asymmetry the rule, and should we then account for symmetry through a form of averaging? Possibly these questions can only be answered as science develops a deeper understanding of the meaning of order—some observations on order are offered in David Bohm and F. David Peat's *Science, Order and Creativity*.

It is also possible to begin with the principle that the universe is not simple, symmetrical, and simply ordered, but rather, at its most fundamental level, its order is one of infinite complexity and flux. Out of this flux would appear orders of relative stability and permanence. The progress of science would be to uncover these orders and to probe their relative limits. Indeed, the history of physics so far could be said to lie in the discovery of what look like fundamental laws and then finding out their limitations and wider contexts. David Bohm has attempted to provide a framework for such an investigation with his notion of the implicate or enfolded order.*

If these ideas seem to be verging on the philosophical, then this is exactly where physics may have to go. Either postmodern physics can continue to be guided by its mathematics, or at some point much deeper insights will be needed. When Einstein was trying to formulate his theory of relativity, he did not study mathematics, nor did he supplement his scientific training by looking at the latest experiments on the speed of light; rather, he read philosophers like Ernst Mach, Immanuel Kant, and David Hume. His essential thought was philosophical, thinking deeply about the meaning of science, the problem of knowledge, and the philosophical implications of space-time. In this way, he was brought to ask some really fundamental questions.

Likewise, when Heisenberg was puzzling over the failures of the old quantum theory, it was Pauli who led him to think in philosophical terms as to whether a new theory should begin with observables or the theory itself should end up by suggesting what is observable. Many of the great revolutions of physics have had at their roots

---

*See, for example, Bohm's *Wholeness and the Implicate Order* (Boston: Routledge and Kegan Paul, 1980).

difficult philosophical considerations. It could well be, as the detractors of superstrings argue, the whole business has moved too quickly and that a great mathematical superstructure has been created over a theory that no one really understands.

On the other hand, it is certainly true that superstrings have given us some very interesting insights and made some intriguing connections. A case in point is the way in which gravitation interactions are pictured in terms of the topological properties of a world cylinder—that is, as the creation and birth of loops. Even if the theory itself does not pay off in the long run, it is very probable that some of its insights will resurface, in totally different contexts, within a grand theory of the twenty-first century. The E8 × E8 symmetry of the heterotic strings, for example, refers to a generalization of symmetries of crystals that were studied many decades earlier. Who would have anticipated that such a mathematical structure, concerned with the patterns of atoms, would ever make contact with a world that lies below the elementary particles? Who would have guessed that twistors, invented to extend Penrose's insights on spin networks, would also end up helping mathematicians solve very difficult nonlinear differential equations?

Physics has always been unpredictable, and it is hard to see where the next advance or insight will come from. It is possible that the theory of superstrings could go the way of the grand unified theories, becoming top-heavy, arbitrary, filled with half-baked and only partly understood ideas. In the end, it would be superseded by some quite different approach. On the other hand, certain aspects of superstrings and twistors could find a new existence in some deeper theory.

Michael Green argues that what is really surprising about string theories is that they work so well. Superstrings have resolved many of the problems that faced earlier attempts to explain the elementary particles; they are free from infinities and associated with just the right symmetry groups. Yet, in Green's opinion, all these theories are fundamentally flawed because they still regard strings as moving in a fixed, background space-time. Such an approach just *has* to be wrong; a proper string theory cannot treat space-time in this way. But the fact that a theory, which cannot at its deepest level be correct, nevertheless appears to work so well may be an interesting clue in itself.

Green's concern reminds me of Conan Doyle's story "The Adventure of Silver Blaze," in which Sherlock Holmes is struck by the fact that the guard dog at the stables did not bark. Rather than focus on something that happened, Holmes was interested in something that did not happen. The failure of the dog to bark meant, of course, that the criminal act had to be done by someone living at the stables, someone familiar to the dog. In a similar way, for Michael Green, the curious thing about superstrings is that they do not bark—they do not cry out that something is profoundly wrong about treating space-time as a background. Somewhere within this mystery is a clue for a latter-day Sherlock Holmes of the theoretical physics world.

It is possible that superstrings may stand in analogy to the atomic theory of Niels Bohr. Following Rutherford's early experiments, and a host of additional experimental data on the spectra of atoms, Bohr attempted to integrate this experimental evidence with Rutherford's atomic model and the new insights of Planck and Einstein on the quantum nature of energy. His solution was a miniature solar system in which the orbits of the planets—the electrons—are quantized into certain discrete orbits. Bohr's "old quantum theory," as it is now called, certainly helped to explain and create order out of a host of experimental data. However, as attempts were made to extend the theory and explain new results more accurately, the whole theory became top-heavy, and arbitrary assumptions had to be made.

Pauli and Heisenberg, for their part, felt that the basic

problem was that Bohr had not gone far enough. He had grafted new ideas—quantization of energy—onto old-fashioned classical ones, like planetary orbits. The old quantum theory was therefore an uneasy mixture of the old and the new, a misshapen hybrid. Could it be that superstrings are the same, that novel ideas are being grafted onto old concepts, pictures, and ways of seeing? Certainly the new superstrings use the same quantum theory that has been around for over half a century; they are formulated in a flat background space-time and rely upon perturbation theory. While everyone recognizes these inadequacies, it is possible there are other paradigms that physics is unconsciously changing and that also need questioning.

In a sense, superstrings are throwing into question the whole Cartesian order by substituting extended objects, the strings, for point coordinates. Yet this revolution in science has happened almost by accident—by extending Nambu's original string model that had been written down in response to Veneziano's attempt to make sense of experimental data. The string revolution really involves a series of fortuitous steps, aided by insight and creative advances. Yet a radical transformation from the Cartesian view of physics really requires a much deeper foundation.

It may well turn out that we need both the brilliant new mathematics and a long and hard look at what superstrings really mean. The time may have come for physics to ask some deep questions, for concealed in one of these may well be the theory of the twenty-first century.

Personal Postscript

AFTER FINISHING THE manuscript of this book, I made a trip to England to meet and talk with some old friends and colleagues: with Roger Penrose and his group at Oxford; Chris Isham at Imperial College, London; David Bohm and Basil Hiley at Birkbeck College, London. I also met, for the first time, Michael Green at Queen Mary College, London.

During my visit, we discussed the merits and the failings of superstrings and of twistors, and on several occasions I began to experience a sense of déjà vu. It seemed to me that after we had explored the particular details of twistor theory or superstrings, we kept coming back to the same questions, questions that I had discussed with exactly the same people almost twenty years ago, questions like: What happens to space-time at the enormously small distances of the Planck length? Does it in fact make any sense to talk about space and time and such distances? What exactly is the meaning of time and its directionality—the "arrow of time"? Has this arrow something to do with the quantum theory, or does it arise at some other level? Can space and time really be considered on an equal footing below the Planck length? Is quantum theory in its final form, or must there be some deeper, underlying theory? Does present-day quantum theory make any sense when we begin to talk about the instant of creation of the universe? What is the deeper meaning of the nonlocal connections that appear to be inherent in the quantum theory? Must quantum theory change as we begin to explore a theory of what could be called prespace? Is it possible to unify quantum theory and general relativity, or must theories emerge from a much deeper form? In fact, do we really understand the quantum theory and the paradoxes it appears to entertain?

As soon as we began to go into these questions, I realized that so little had changed, that I had explored such questions almost twenty years ago, and with exactly the same people. Despite the advances of superstrings, despite two decades of work on quantum gravity, despite the many new insights into twistors, in a fundamental way these questions still remained unanswered. Chris Isham, for example, points out that by assuming the validity of quantum theory, we are free to play around with new ideas about space-time structure. We can explore different topologies, attempt various algebraic relationships, but all the time assuming that quantum theory holds. On the other hand, we can ask how gravity and space-time geometry will change quantum theory—that is, attempt to transform the quantum theory from within the context of general relativity.

But what on earth can we do once we admit that both theories may change, that quantum theory and general relativity are not the final answer? Once we entertain the possibility of dropping both theories there seems to be nothing to hold on to, no deep guiding principle that would suggest a path through this wilderness. In Isham's opinion, the problems that face physics today are far more difficult that those which Einstein tackled at the start of the century when he transformed Newtonian mechanics in order to reconcile the ideas of space-time with Maxwell's theory of electrodynamics. Einstein had a strong foundation on which to stand, yet today our foundations are uncertain; we do not know where to place our feet.

Despite its triumphs, superstring theory seems to have been unable to answer these great questions. But, for me, one of the most exciting outcomes of the theory is that it is finally bringing so many talented physicists face-to-face with some very deep questions. In the past it was possible to skirt past these issues and avoid them, but now, if string theories are to make real progress, it will be necessary to face these issues head on.

In the first chapter of this book, I discussed a potential revolution in physics, one in which the Cartesian ideas of continuity and dimensionless points would be replaced by new ways of thinking that are more in harmony with the insights of the quantum theory. String theory looked like a major part of this revolution. Nevertheless, so much of the theory, so much of its thinking still clings to old, Cartesian ways of thinking. Now, it may well be that in some physics of the twenty-first century, while quantum theory and relativity are transformed and a Cartesian order is thrown away, some valuable aspects of, shall we say, Cartesian mathematics will remain. But it is hard to believe that the old Cartesian order can remain in the face of all the implications of the quantum theory.

Up to now the mainstream of string theory has failed to displace the Cartesian order. But what of twistor theory? This approach is certainly consistent in its search for a new order, one that begins globally rather than locally, one that emphasizes not points but global structures. Within the twistor approach there is no need to hang onto a smooth, continuous space-time, for all indications are that space-time and the quantum theory both will be transformed by twistors. On the other hand, while twistor theorists are committed to a truly revolutionary approach, the whole theory has proved far more difficult than anyone anticipated, and the initial promise of the theory has not yet been fulfilled. While proponents of twistors have made advances in a number of fields, some of the deepest problems still call for new ideas. The same could be said for some of the more innovative superstring approaches that are attempting to get away from the idea of space-time as a background. While such ideas may be interesting, again they have yet to be worked out in their full form.

Of course, superstrings and twistors do not represent the only approaches that are attempting to go beyond the Cartesian order or that seek to transform quantum theory and the general theory of relativity. On the one hand, there are a number of mathematical excursions into new forms of description. But often, while being speculative, they do not have a firm philosophical underpinning or a compelling physical motivation. At times they almost seem like shots in the dark. (Of course, it is possible that one of these shots will hit the target, and then physicists will be faced with the major question of just why.)

Then there are the more philosophical approaches, and here I have in mind David Bohm's notion of the implicate order. Bohm's ideas are well argued, and it is convincing that a new order is required by quantum theory, an order that is essentially nonlocal and of an enfolded, rather than an explicate, nature. Certainly a new order, which may well have many of the characteristics that Bohm suggests, is demanded by any theory that attempts to transcend quantum theory. But these ideas of Bohm have still to excite the interest of most theoretical physicists. The problem is that such an order does not yet have a mathematical form, and needs to be translated into formal relationships that could replace the more conventional treatment of space and time.

On the one hand, there are mathematical excursions with no deep foundation; on the other, there are ideas for new approaches that have not evolved an explicit mathematics. There is the promise of twistor theory, which has yet to be fully worked out. There is the juggernaut of superstring theory, which can no longer continue on its present course unless some challenging issues are faced. And so the deepest questions remain. But at least more and more physicists are realizing that a crisis does indeed exist in physics, that hard work is required and profound new ideas are called for.

So this book ends with unanswered questions. They are the same questions I first began to ask myself as I learned about quantum theory as a student. They are probably the same questions that you will have asked when you began to read about quantum theory and relativity in popular science books and magazines. They are the same questions that physicists return to again and again. In past decades, however, it was possible to avoid these questions, to forget about them and get on with the mainstream business of physics. But today, for better or for worse, superstring theory has got itself involved in asking some very provocative questions about the nature of space and time at enormously small distances and, at the same time, realizing that some profound new ideas are called for.

It seems to me that these outstanding questions can no longer be avoided. Someone will have to answer them. I hope that when they do, I'll still be around to write about them!

Suggested Additional Reading

TO LEARN MORE about elementary particle theory, quarks, gauge theory, grand unification, and supersymmetry the general reader is referred to the very many books that present this branch of physics in a nontechnical and nonmathematical way. Such books are readily available in libraries and book stores, and it would be difficult to recommend one over the other. Articles that go into particular issues in even greater depth can be found in the pages of *Scientific American*. Since such articles are written by acknowledged experts in the particular field they have the stamp of authority. However, *Scientific American* articles are not always easy to follow at the first sitting and the reader must persist, rereading the same passage several times and referring to the diagrams that are often very helpful.

When it comes to the topics of Superstrings and Twistors it is not so easy to suggest additional reading. They are highly technical and mathematical topics and the researchers involved have not yet found time to write more popular articles and reviews. A noted exception is Michael Green's *Scientific American* article on superstrings, which can be found on page 48 of the September 1986 issue (Volume 255, Number 3). Other popular overviews can be found in *Science*, 1985 (Volume 229) page 1251 by M. M. Waldrop and in *Discover*, November 1986 (Volume 7, Number 11) page 34 by Gary Taubes.

Physics students are told how to quantize a classical string by Guy Fogleman in *American Journal of Physics* (Volume 55, Number 4) April 1987. Graduate students may be interested in Joel Scherk's account of the dual resonance model and early string theory in *Reviews of Modern Physics* (Volume 47, Number 1) January 1975, page 123. For the technically minded, the standard book on the topic is the two volume *Superstring Theory* by M. B. Green, J. H. Schwarz, and E. Witten published by Cambridge University Press in 1987. While the treatment is mathematical, there are helpful summaries at the end of most of the chapters. In addition, Chapters 12 and 15, (which are found in Volume 2), contain valuable introductions to the mathematics of algebraic and differential geometry—techniques that are used in both superstring and twistor theories. The book contains an excellent bibliography referring the reader to all of the important papers on the topic—there is no point in reproducing that list here. Although the more recent advances of string field theory are not treated in this book, a series of references to key papers, published up to 1986, is included.

David Gross has written a technical but nonmathematical review of his Heterotic String Theory in *Physica Scripta*, Volume T15 (1987) page 34. The giants of string theory also give lectures containing overviews of the field at conferences and seminars. Some of these lectures are reprinted in the various conference proceedings that can be found in university libraries. These general articles are usually placed at the beginning of the conference proceedings and act as a useful orientation for the physicist or mathematician who may not be up-to-date about the latest details in that particular field.

When it comes to twistors and spinors the standard book is the two volume *Spinors and Space-Time* by R. Penrose and W. Rindler, published by Cambridge University Press. A much shorter, and less expensive introduction, *Advances in Twistor Theory* has been written by L. P. Hughston and R. S. Ward and is published by Pitman,

London, 1979. But again, these books are highly technical and require the mastery of some modern mathematics. Penrose has written a short nonmathematical history of the development of his ideas, "On the Origins of Twistor Theory," which appears in *Gravitation and Geometry*, edited by W. Rindler and A. Trautman, published by Bibliopolis, Naples, 1987. For the technically minded, Penrose's key papers are "Angular Momentum: An Approach to Combinatorial Space-Time" published in *Quantum Theory and Beyond*, edited by T. Bastin, Cambridge University Press, 1971; "Twistor Algebra" in *Journal of Mathematical Physics*, Volume 8 (1967) page 345; "On the Nature of Quantum Geometry" in *Magic without Magic: J. A. Wheeler* edited by J. R. Klander, Freeman, 1972; "Twistor Quantization and Curved Space-Time," in *Internal Journal of Theoretical Physics*, Volume 1, Number 1 (1968) page 61; and "Nonlinear Gravitons and Curved Twistor Theory" in *General Relativity and Gravitation*, Volume 7 (1976) page 31. Twistor specialists communicate via the *Twistor Newsletter*, which is privately circulated from the Mathematics Institute of Oxford University. Penrose himself is currently at work on a nonmathematical book, which appears to go into a number of very exciting areas such as the meaning of quantum theory and space-time as well as his speculations on the nature of consciousness. This book is eagerly awaited by all of us.

As to the types of mathematics that are used by the new generation of physicists, a nontechnical introduction that refers to some of these techniques can be found in *Concepts of Modern Mathematics* by Ian Stewart, published by Penguin Books. The edition I have on my book shelves, with a new appendix, is dated 1981. Technical readers who already have a background in mathematics are referred to Robert Geroch's *Mathematical Physics*, published by The University of Chicago Press, 1985. This book is based on a university lecture course and contains the sorts of mathematics that are currently being used in relativity and particle physics. As already mentioned, topics such as differential and algebraic geometry, cohomology and the properties of Calibi-Yau spaces are treated in Chapters 12 and 13 (Volume 2) of Green, Schwarz, and Witten's *Superstring Theory*. The two volume set on *Spinors and Space-Time* by Penrose and Rindler also explains some of the mathematics involved.

Glossary

action: A quantity related to the momentum and position of a body or system of particles. The Principle of Least Action asserts that the integral, or sum of this action, taken over a particular path must be a minimum. This Principle of Least Action can be used instead of Newton's Laws to determine the motion of a system.

bosons: Elementary particles that have integral spins. Force particles such as the photon, gluon, and vector bosons are all bosons. But note that there can also exist composite particles formed out of collections of fermions—such as a helium atom—which act collectively as bosons.

bootstrapping: A hypothesis about the nature of the elementary particles derived from S-matrix theory. It suggests that no particular particles are more elementary than any others but that they all arise out of each other in a democratic and self-consistent fashion. That is, the elementary particles pull themselves up by their own bootstraps.

Calibi-Yau space: These six-dimensional spaces are hypothesized as arising when the ten dimensions of superstring theory are compactified down to four dimensions. They are also related to orbifold spaces.

chirality: An expression of the basic handedness of nature. Fundamental theories of the elementary particles and of superstrings must possess chirality.

color: A quantum number or characteristic of quarks.

compactification: The process of "curling up" six of the ten dimensions of superstring theory.

conformal geometry: Conformal geometry is related to the stretchings of space-time that preserve the light cone structure. A space containing only null lines, such as Penrose's basic twistor theory, is conformally invariant. Both mass and nonlinear interactions, such as gravity, break conformal invariance.

contour integral: A powerful mathematical tool used in complex geometry whereby the value of an integral is determined by drawing a contour or boundary and evaluating singularities, poles, and residues inside.

Cohomology: A branch of mathematics concerned with the patching together of spaces.

cosmic strings: Some contemporary cosmological theories suggest that boundaries were formed between different regions of the universe at the moment of creation. These boundaries survive today as "cosmic strings," incredibly thin but very massive strings many light years in length.

cosmological constant: Einstein's general theory of relativity allows for space-time curvature even in an empty universe. The amount of this curvature is given by the cosmological constant. Current indications are that this constant must be zero, but the reason for its vanishing remains a mystery.

covariance: According to the general theory of relativity, theories of nature must have the same mathematical form in all coordinate systems. A good physical theory must therefore be written in a covariant form.

dispersion relations: Formal relationships between the real and the imaginary parts of a complex mathematical function. In physical terms dispersion relations have been used to relate terms in the scattering matrix for elementary particle theories.

Dual Resonance Theory: A theoretical model that was developed to explain some of the experimental properties of the scattering matrix, such as the duality between s- and t-channel scattering. It was Veneziano's version of the dual resonance model that paved the way for Nambu's string theory.

electroweak force: A unification of electromagnetism and the weak nuclear force.

fermions: Elementary particles with fractional spins. The proton, electron, neutron, and other elementary particles are all fermions.

Feynman diagrams: Used to calculate the forces and interactions between elementary particles. Feynman diagrams are also of importance in calculating the S-matrix. Related diagrams are found in both twistor theory and superstring theory (the trouser diagrams).

Gauge Theory: A theory that treats force in a geometrical way in terms of global and local symmetries.

gluon: The hypothetical particle that carries the force between quarks.

hadrons: Strongly interacting elementary particles. The hadrons all tend to have high masses.

helicity: This could be thought of as the "spin" of a massless object such as a twistor, null line, or photon of light.

heterotic string: Gross's version of string theory in which space-times of different dimensions are associated with the same closed loop.

homogeneity: This expresses the power of a mathematical function. Thus x^3 has homogeneity of +3, while $\frac{1}{x^4}$ has homogeneity of −4. The homogeneity of twistor functions is of key importance in expressing fields in twistor terms.

implicate order: A term coined by the physicist David Bohm to describe the sort of enfolded order that is characteristic of quantum theory. It is to be contrasted with the explicate orders of Newtonian physics. Bohm believes that this implicate order has a universal importance and will be useful in understanding the nature of consciousness.

internal symmetry: The properties of different elementary particles can be related to each other by mathematical transformations that look very much like the more familiar symmetry properties of our own physical space. Physicists have therefore hypothesized an abstract internal space in which these internal symmetries are defined. With the help of these internal symmetries, the elementary particles can be gathered into families. The relationship between space-time and these internal symmetries remains to be fully explained.

isospin: The neutron and proton behave as if they have an additional form of spin that transforms one into the other. This spin cannot, however, exist in our physical space and is referred to an abstract space—isospace.

Kaluza-Klein Theory: An early attempt to unify general relativity and electromagnetism by working in five dimensions. The electromagnetic field was obtained by curling up or compactifying the extra dimension. With the advent of higher dimensional theories such as superstrings, the Kaluza-Klein approach came back into fashion.

grand unification: An attempt to produce a unification of all the forces of nature. While some success was made in unifying the gluon force between quarks with the electroweak force, problems always arose when gravity was included. Grand unification eventually gave way to superstring theory.

graviton: The hypothetical quantum particle of the gravitational field. It could also be thought of as a quantized element of space-time curvature.

least action: *See* Action.

leptons: Elementary particles like the electron and neutrino that do not experience the strong nuclear force. Unlike the strongly interacting hadrons, the leptons have small masses.

light cone: Light emanating from a source forms, in three-dimensional space, a sphere that expands in time. In space-time, however, it produces a conical structure. Points within this light cone can be causally connected together. The light cone structure is of key importance in relativity. The light cone is unchanged by conformal transformations.

null line: The path, in space-time, of a light ray or other massless object. Space-time distances measured along a null line are zero.

minimal surface: A mathematical term referring to surfaces that satisfy a minimization procedure. Soap bubbles, for example, minimize their energy by forming shapes with the minimum possible area. The world surface of a string is likewise a minimal surface.

monopole: A hypothetical quantum object being a single, isolated magnetic pole. Normally, magnetic poles, the sources of a magnetic field, occur in pairs as north and south poles.

orbifold: A particular space used as a candidate for the compactified space of superstring theory. These six-dimensional orbifolds could be thought of as generalizations of a six-dimensional torus, but containing twenty-seven singular points.

Planck length: The size limit at which normal notions of space-time are supposed to break down.

quantum gravity: A general term used to describe attempts to quantize gravity. The elementary particle of the gravitational field is the graviton.

quark: The hypothetical constituent of the elementary particles that interacts via glue forces. Originally only three quarks were hypothesized; today it appears that six are required. For a variety of theoretical reasons, free quarks can never be seen.

Regge Trajectory: Derived from S-matrix theory, the Regge Trajectories were theoretical plots that attempted to account for the position of elementary particle resonances. One of the triumphs of early string theory was to describe the general shape of these Regge Trajectories.

resonance (elementary particle): Resonances are sometimes found when elementary particles collide and interact together. They represent tiny regions of space in which energy is temporarily bound.

torus: The topological name for the shape of a donut. While a donut is a two-dimensional surface in a three-dimensional space, the torus can be generalized to higher numbers of dimensions.

scattering matrix/S-matrix: The S-matrix relates the incoming and outgoing states of elementary particles during interactions and scattering experiments. The mathematical structure and properties of the S-matrix has received considerable attention.

second quantization: This goes beyond the quantum theory of Heisenberg and Schrödinger by applying the act of quantization a second time. In this way, matter and energy fields can themselves become quantized. The quantum excitations of these fields are the elementary particles.

spin network: A term used by Roger Penrose to denote collections or networks of quantum mechanical spinors. Although they were not created within any background space, Penrose discovered that these spin networks had properties that were similar to those of Euclidian angles in a three-dimensional space. One of Penrose's early goals was to extend the spin network idea by employing twistors and in this way derive the properties of the space-time quantum mechanically.

spinor: A quantum object that describes the basic two valuedness of the electron, proton, neutron, and other spin half elementary particles. The mathematical spinor also has a role to play in relativity theory since it can be used to define the light cone.

strangeness: A quantum number used in quark theory.

string: Nambu's original idea that the elementary particles could be described as extended, one-dimensional objects was called string theory. Since the ends of Nambu's strings whipped around at the speed of light they were also called light strings. Later attempts to include the spin half fermions within a string theory led to the term spinning strings. Strings that possess supersymmetry are called superstrings. Heterotic strings combine spaces of two different dimensionalities. The term string is used in a generic way to describe all these different variations, including superstrings.

supersymmetry: A symmetry that relates the fermions (fractional spin particles) to the bosons (elementary particles with integral spin).

symmetry (elementary particle): Abstract mathematical relationships that relate elementary particles together and allow them to be grouped into families. A particular symmetry transformation has the effect of, in a theoretical way, transforming one elementary particle into another. Important symmetries include:

U(1): the symmetry of the electromagnetic field.

SU(2): the symmetry of the weak nuclear interaction.

SU(2) × U(1): the symmetry of the unified electroweak interaction.

SU(3): the symmetry corresponding to quark theory and the strong nuclear interaction.

SU(5): One of the suggested symmetries of the grand unified theory in which the gluon and electroweak forces are united. It includes the group SU(3) × SU(2) × U(1).

twistor: The twistor is a sort of generalization of a spinor, being a massless object having both linear and angular momentum. It can be defined in terms of a pair of spinors. Twistors are the coordinates of twistor space, but they also have a geometrical interpretation in space-time. Twistors with zero helicity correspond to null lines while more general twistors must be pictured as congruences of null lines.

vector boson: Force-carrying particles of nature. Three vector bosons are responsible for the weak nuclear force. By admitting the photon on an equal footing it is possible to create a unified electroweak theory. As a result of symmetry-breaking processes, however, this photon remains massless while the three other vector bosons pick up mass.

Veneziano Theory: A formula that accounted for the experimental results of the dual resonance model. Nambu discovered that a string theory would reproduce the results of the Veneziano approach.

virtual particle/virtual interaction: Quantum uncertainties in energy make it possible for virtual particles to be constantly created and annihilated during elementary particle interactions. Elementary particles are able to make use of these virtual particles within their interactions.

world line: The path in space-time traced by a body.

world surface/world sheet: The surface traced in space-time by an extended object such as a string.

Index

Action, 102–3, 110
Aging of universe, 322–24
Alpha particles, 31
Alvarez-Gaume, 115–17, 277
Ambitwistor space, 309
Anderson, Carl, 75
Angular momentum, 182
Anomalies, 116–18, 133, 134, 277, 286
Antiparticles, 75
Artificial intelligence, 28
Asymptomatic freedom, 82

Bach, J. S., 276, 280
Baryon number conservation, 90–91
Big bang, 149, 210, 225, 277
Black holes, 19, 20, 127, 129, 210, 225
Blobs, and resonance theory, 44–45, 49, 57
Bohm, David, 25–26, 27, 45, 333, 335, 339, 342
Bohr, Niels, 80, 209, 337, 338
Bootstrap theory (self-referential particles), 43
Bowick, M. J., 300, 311–12
Briggs, John, 259

Calabi-Yau space, 156–61, 299–302
Calculus, 19, 22, 24, 241. *See also* Differential equations
Cartesian coordinates, 22–23, 25, 103–5, 188
Cartesian order, 338, 341, 342
Chew, Geoffery, 43
Chirality (handedness), 111–13, 114, 116, 118, 134, 205, 214, 231, 249–54, 265–66, 286, 293, 297
Clifford, William Kingdon, 201–2, 204, 246, 268, 308
Closed-loop theory, 114–19
Coherence, 279, 308
Cohomology, 165, 180, 183–85, 215, 238, 241, 243, 245, 268, 306, 334
  and Penrose triangle, 183–85
  sheaf, 249
Color, 51, 77, 82
Compactification, 113, 116, 117, 120, 130–31, 137, 139, 148, 154, 156, 158, 214, 277, 285–88, 289–91, 296, 298, 299, 300–2, 311, 333
Complex dimensions, 61, 204

and twistor space, 212, 214
Complex numbers, 164, 183, 185–87, 189, 195, 197
Complex space, 165, 198, 215
Complex worlds, 183–91
Conformal invariance, 215–18, 226, 230
and null lines, 195–96
Congruence, 311
Conical singularities, 291–92
Consistency, internal, 277, 279
Continuity, 21, 24, 174–75
Contour integrals, 241–47, 262–65
Coordinates. *See* Cartesian coordinates
Cosmic strings, 283, 324–28
Cosmological constant (zero), 17, 154, 330
Covariance, 104, 108, 281. *See also* Manifest covariance
cummings, e. e., 116
Curvature of space-time, 13, 154, 170, 266, 284, 302, 313
and spin networks, 180
and twistors, 218–23

da Vinci, Leonardo, 102
de Broglie, Louis, 45, 332
Descartes, Rene, 22–24, 308. *See also* Cartesian coordinates
Diatonic scale, 52, 54
Differential equations, 19, 241. *See also* Calculus
definition of, 24
Differential field equations, 239, 241, 247, 257–59, 265
Dimensionality, of space-time, 16–17, 64, 67, 68, 74, 120, 136, 152–56, 203, 231, 232, 283–84, 286, 288, 290, 298, 300, 307, 319–21
and spin networks, 178–79
and superstrings, 113–14, 293–94
Dimensionless points, 10, 12
Dirac, P. A. M., 75, 304, 322–23
Disintegration, 112
Dispersion relations, 42–43, 189
Donaldson, S., 313
Doyle, Arthur Conan, 337
Dual resonance, 45–51, 74, 186
and harmonics, 54–55
and string theory, 58–61

Eightfold way (symmetry), 78–80
Einstein, Albert, 72, 73, 103–9, 126, 147, 176, 202–3, 210, 255–59, 265, 266, 303, 312–13, 324, 332, 335, 337, 340. *See also* Relativity
and cosmological constant, 17
and gravity, 272
and space continuity, 19, 21–22
and space-time curvature, 17, 20
and theory of energy, 37
Elementary particles
as one-dimensional, 28
as self-referential, 43
definition of, 11–13
and energy, 11–12
families of, 33–35, 48
interaction of, 115
and quarks, 35
resonances, 32
spin of, 27
and symmetry, 11–12, 15–16
tachyons, 64–65
and universal force, 90
Energy scales, 96
Enfolded order. *See* Implicate order
Escher, M. C., 165, 183
Euclid, 201, 231
Euler characteristics (numbers), 157, 158–59
Euler, Leonhard, 102, 157

Event horizon, 127, 129

Faraday, Michael, 242-43
Feynman diagrams, 38-42, 122, 128, 145, 230, 302
Feynman, Richard, 38, 330
Feynman series, 64
Fiber bundles, 267, 272, 309
Field theory, 304-5, 313
Finkelstein, David, 28
Fishbach, Ephraim, 324
Fivefold symmetry, 166-68
Four dimensions, 16, 49, 60, 106, 130
Friedan, Daniel, 310-11

Galileo, 256, 324
Gauge fields, 84-86, 88-89, 97, 115, 122, 202, 267-68, 284
Gauge forces, 89, 114, 120-21
Gauge invariance, 83
Gauge theory, 83-91
Gell-Mann, Murray, 65, 78-80
Geodesics, 257
Ghosts, 47-48, 60, 64, 68, 109
Gilbert, William, 120
Gliozzi, F., 68
Global symmetry, 84-85
Glue force, 82, 89
Gluon force, 94, 120
Goudsmit, Samuel, 75-76
Grand unification theory, 9-10, 14, 90-91, 93-94, 97, 98, 285, 286, 336
Grand unified symmetry, 118, 287, 292, 293-94, 300, 305-6
Grassmann, Hermann Gunther, 308
Gravity, 13-14
  and grand unified theory, 97, 169
  and quantization, 259-67
  and space-time curvature, 18
  and string interaction, 63
  and superstrings, 123-28
Green, Michael, 64, 65, 68, 98-101, 110, 113, 114-21, 126, 128-31, 133-34, 140, 161, 277, 286, 288, 304, 316, 318, 319, 320, 336-37, 339. *See also* Schwarz, John. Superstring theory, Superstrings
Gross, David, 134-37, 147, 161, 286, 288, 292, 304, 306, 316, 319, 320. *See also* Heterotic string theory

Hamilton, William Rowan, 102
Handedness. *See* Chirality
Harmony, 51-56
Harvey, Jeffrey, 134
Hawking, Stephen, 19
Heisenberg, Werner, 32, 76, 190, 197, 237, 304, 335, 337
Heisenberg's uncertainty principle, 20-21, 25, 40, 219, 226
Helicity, 205, 208, 211, 233, 248, 250-53, 263, 265-66
Hero of Alexandria, 102
Heterotic string theory, 133-62, 147, 154, 286-87, 292, 306, 316, 320, 336. *See also* Gross, David
Hiley, Basil, 339
Hodges, Andrew, 230, 232, 234, 285
Holomorphic curve, 281-83, 327
Homogeneity, 248, 251-53, 264-65, 309
Hughston, Lane, 214, 226-27, 273, 281, 283, 327
Hume, David, 335

Implicate (enfolded) order, 26-27
Inertial mass, 256
Infinite divisibility, 24

Infinities, 64-65, 68, 109, 134, 225, 277, 286
and elementary particles, 11-12
and interactions, 145-47
Initial grand symmetry (E8 × E8), 137-38, 145, 147, 148-51, 157, 161
Interactions, 140, 145-47
Interference, 171-74
Internal space, 77
Internal symmetry, 101
Isham, Chris, 339, 340
Isospace, 77
Isospin, 77, 78, 80, 92

Kahler geometry, 311
Kahler manifold, 300-1
Kahler space, 311-12
Kaluza, Theodor, 115, 203. *See also* Kaluza-Klein theory
Kaluza-Klein theory, 115-16, 137, 331
Kant, Immanuel, 335
Kepler, Johannes, 53
Kibble, T. W. B., 326
Klein, Oskar, 115, 203. *See also* Kaluza-Klein theory
Kolbe, Edward W., 149
Kuhn, Thomas, 133

Lagrange, Joseph-Louis, 102
Lamb shift, 40
Lande, 332
Least action, principle of, 102, 282
Lee, Tsung-Dao, 112
Leibniz, Gottfried, 102
Leibniz, Wilhelm, 171
Light cone, 195, 215-18, 223-24, 265
Light cone gauge, 314-16
Light rays. *See* Null lines
Light string. *See* String theory
Lines of force, 242-43
Lobachevsky, Nikolay Ivanovich, 201
Logical postivism, 176
Lovelace, Claude, 60

Mach, Ernst, 176, 335
Mandelbrot, Benoit, 321
Mandelbrot set, 188
Mandelstam, Stanley, 128
Manifest covariance, 57, 118, 313, 314, 315. *See also* Covariance
Manifolds, 300-1
Martinec, Emil, 134
Mass, problem of, 292-99
Maxwell, James Clerk, 72, 120, 242, 248, 272, 340
Mendeleyev, Dmitry, 78-79
Minimal surface, 281-83
Minkowski, Hermann, 105-6, 191, 203, 207, 259, 288
Missing mass, problem of, 148
Music of the spheres, 53

Nambu, Yoichiro, 49, 51, 55, 65, 74, 99, 100, 110, 135, 229, 276-77, 319, 320, 328, 338. *See also* String theory
Ne'eman, Yuval, 78-79
Neilson, H., 53
Neural nets, 28
Neveu, Andre, 67
Newton, Isaac, 21, 22, 280
Newtonian mechanics, 19, 21, 40, 53, 72-73, 101, 103, 110, 171, 177, 238-39, 258-59, 314, 340
Nonlinearity, 259-60
Nonlocality, 168-69, 209-10, 225, 269-70
Null lines, 191-96, 204, 207-9, 211-12, 213, 215, 222, 226, 233, 265, 311
and conformal invariance, 195-96

and geometry, 197-200
and speed of light, 195

Oliver, D., 68
Omega minus particle, 78-80
Operators, 109-10
Orbifolds, 156-61, 288-92, 299, 302

Particle masses, 151-52
Pauli, Wolfgang, 112, 335, 337
Peat, F. David, 27, 335
Penrose, Lionel, 165
Penrose, Roger, 10, 13, 16, 27, 28, 49, 61, 76, 111, 162-200, 203-4, 215-18, 221, 223, 227-32, 234, 237-41, 247, 252, 254, 255, 261-62, 265-67, 269-72, 283-85, 297, 303, 309- 11, 314, 318, 319, 334, 339. *See also* Twistors
Penrose, Sir Roland, 165
Penrose triangle, 165-66, 183-85, 238
Periodic table, 78
Perjes, Zoltan, 227
Perturbation theory, 39, 115, 259-61, 280-81, 311, 316
and superstrings, 302-3
Planck, 337
Planck length, 312-13, 318-19, 329, 339
Plenum, 22
Podolsky, Boris, 210
Polarization, 249-51
Postmodern physics, 275-80
Poylakov, A. M., 128
Projective geometry, 164, 204
Pythagoras, 51-53, 191-92

Quantization, 197, 269
and gravity, 259-67
and relativity, 260
second, 101, 305
and superstrings, 109-11
Quantized force, 122
Quantum chromodynamics theory, 82
Quantum electrodynamics, 40, 82
Quantum field theory, 24-25, 101, 120, 197, 225, 237, 304, 322
Quantum fluctuations, 223-26, 287
Quantum geometry, 170-74
Quantum theory, 10-11, 75, 102, 110, 258, 270, 340
and chirality, 112
and compactification, 287-88
and continuity, 21
and electromagnetism, 14
and elementary particles, 12
and helicity, 252
and light cones, 216
and nonlocality, 168-69, 209-10
and nuclear forces, 14
and relativity, 11, 17-19
and space-time curvature, 13
and string theory, 58-60
Quark theory, 35, 49
Quasicrystals, 166-68, 269

Rajeev, Sarada, 300, 311-12
Ramond, Pierre, 66-67
Regge graphs, 229
Regge poles, 47
Regge trajectories, 43-45, 55, 59, 227, 228
Regge, Tulio, 43-44
Relativity, 10-11, 73, 127. *See also* Einstein
and chirality, 112
and compactification, 287-88
and infinities, 147
and interactions, 147
and light cone gauge, 314
and mass, 228
and null lines, 191, 193

and quantization, 260
and quantum theory, 11, 17–19, 169–70
and space-time curvature, 13
and string theory, 57–58
and superstrings, 103–9
and twistor space, 207
Resonance, 32, 43–51, 78. *See also* Dual resonance
Riemann, Bernhard, 201, 311–12
Robinson congruence (congress), 199, 200
Robinson, Ivor, 198
Rohm, Ryan, 134
Rosen, Nathan, 210
Rubbia, Carlo, 89
Rutherford, Ernest, 31, 35–37, 275, 337

s-channel scattering, 46–47, 50–51
S-matrix (scattering matrix), 37–42, 45, 186, 189
Salam, Abdus, 86, 320
Scalar field, 267
Scherk, Joel, 67, 68, 92
Schrodinger, Erwin, 21, 190, 197, 237, 304–5, 332
Schrodinger's wave equation. *See* Wave equation
Schwarz, John, 64, 65, 67, 98–101, 110, 113–21, 128–31, 133–34, 140, 161, 277, 286, 288, 304, 316, 319, 320, 328. *See also* Green, Michael, Superstring theory, Superstrings
Seckel, David, 149
Shadow universe, 147–51
Shaw, William T., 214, 281, 283, 327
Sheaf cohomology, 249. *See also* Cohomology
Shenker, Stephen, 310–11
Singer, Michael, 214, 232, 234, 285
Singularities, 210, 225, 244–245, 249, 264
conical, 291–92
Space-time
dimensionality of. *See* Dimensionality
foamlike structure of, 28, 170–71, 226, 318
lattice structure of, 28
theory of, 105
Sparling, George, 272
Spin, 93, 99, 112, 125, 164, 176–77, 309
classifications, 33
integral, 66
isospin, 77, 78
quantized, 76
and resonances, 44–45
and string interaction, 63
Spin networks, 28, 174–82, 197, 231, 319
and curvature, 180
and twistors, 191
Spinning string theory, 67–68, 76
Spinor networks, 200
Spinors, 27, 67, 76, 177–78, 224, 229
and light cones, 216
complex, 195
definition of, 175
Strangeness, 77, 78, 80, 92
String field theory, 308, 311–12
String fields, 303–14
String interaction, 62–64, 121
String theory, 49, 51, 57–61, 68, 74, 99, 152, 163, 228, 232–33. *See also* Nambu, Yoichiro
and dual resonance, 58–61
and free quarks, 81
and grand unification, 98
and harmonics, 54–56, 57

and infinities, 64–65
and mass, 58
number of theories, 316–19
one-dimensionality, 61
and quantum theory, 58–60
and relativity, 57–58
and supersymmetry, 92, 98
Strings
bosonic and fermionic, 66–68
cosmic, 283, 324–28
length of, 59, 99
and quarks, 65–66
Super wave function, 306. *See also* Wave function
Supergravity, 10, 68, 93, 99, 115. *See also* Gravity
Superspace, 97
Superstring theory, 10, 276, 286
and mass, 296–97
and phenomenology, 299
problems with, 328–33
Superstrings, 11, 99, 163, 310
and aging of universe, 322–24
and closed loops, 114–19
and cosmic strings, 327–28
definition of, 100–11
and dimensionality, 113/-14, 293–94
and elementary particles, 12–13
and forces, 120–21
and gravity, 14, 123–28
internal symmetry of, 101
and perturbation theory, 302–3
postmodern, 280–81
and quantization, 109–11
and relativity, 103-9
and supersymmetry, 34
and twistors, 213–15, 281–85
and vibration, 101–3
Supersymmetry, 10, 28, 67–68, 69, 91–95,100, 118, 285, 286, 305. *See also* Symmetry
and chirality, 112, 114
and local symmetry, 93
and string theory, 98
and superstring theory, 34
Susskind, Leonard, 53
Symbolic logic, 28
Symmetry, 11–12, 15–16, 78, 122, 138, 293
eightfold way, 78–80
global, 84–85
local, 84–85
of physics, 75
patterns, 97
time reversal, 111
Symmetry breaking, 15–16, 87–88, 111, 148, 149, 152, 155–56, 214, 252, 277, 285, 294, 297, 324, 334
Symmetry groups, 11, 78, 87, 118, 133

t-channel scattering, 46–47, 50–51
Tachyons, 64–65, 68, 74, 109
Temporal logic, 28
Tetrakyts (fourness), 51–52
Theory of everything, 13, 98, 118, 119, 328
Thermodynamics, 72
Topological manifold, 300–1
Topology, 25–27, 50–51, 122, 124, 145, 153, 157, 158, 308, 336, 340
Trouser diagram, 141–45, 232–34, 253, 315. *See also* Twistor diagram
Tsou, Sheung Tsun, 229
Turner, Michael S., 149
Twistor diagram, 285, 229–35. *See also* Trouser diagram
and gravity, 254
and twistor networks, 229–35
Twistor fields, 237–42
Twistor gravity, 237–73

Twistor space, 196, 199, 201–35, 254, 262–65, 267–68, 281, 285, 300, 309
  and complex dimensions, 212
  and relativity, 207
Twistor theory, 10, 196. *See also* Twistors
Twistors, 10, 11, 27, 29, 76, 204, 277. *See also* Penrose, Roger
  and contour integrals, 246–49, 252–53
  and curvature, 218–23
  and elementary particles, 12–13
  and geometry, 200
  and gravity, 14
  one-dimensionality of, 61
  origins of, 163, 182–83
  and particles, 226–29
  and spin networks, 191
  and spinors, 163–200, 205
  and superstrings, 213–15, 281–85
Two-valuedness, 75–76
Type I theory, 100, 114, 117, 119, 134
Type II theory, 100, 114, 117, 134

Uhlenbeck, George, 75–76
Ultimon, quest for, 12
Uncertainty principle. *See* Heisenberg's uncertainty principle
Unification theory, 19–21, 28, 71–72, 258
Unified field theory, 86, 237
Unified force, 90
Universe, shadow. *See* Shadow universe
Unreasonable effectiveness of mathematics, 52–53

Vector space, 268
Veneziano, Gabriele, 45–47, 59, 338
Vigier, Jean-Pierre, 45
Virtual interactions, 40
Virtual particles, 40
Voltaire, 102
von Eotvos, Roland, 256, 324
von Weisacker, Karl, 28

Ward, J. C., 86–87
Ward, R. S., 267–68
Wave equation, 21, 24, 269–70
Wave function, 101, 110, 190, 206–7, 254, 263, 304–5, 311
  collapse of, 269–72
Weinberg, Steven, 86–87, 320
Weyl, Hermann, 202
Wheeler, John, 37–38, 128, 170, 201, 210, 226, 238
Wigner, Eugene, 52–53
Witten, Edward, 113–14, 115–17, 138–39, 156, 162, 183, 215, 269, 277, 281, 283–84, 303, 306–9, 312–13, 327
World lines, 106, 224–25
World surface, 107, 108, 123
Wu, C. S., 112

Yang, Chen Ning, 112
Yang-Mills forces, 114, 247, 269
Yau, S. T., 300

Zel'dovich, Y. B., 327
Zero, as cosomological constant, 17
Zumino, Bruno, 169
Zweig, George, 65, 80